Spectral Problems
in
Organic Chemistry

R. DAVIS and C.H.J. WELLS

Kingston Polytechnic, Kingston-upon-Thames, England

International Textbook Company

Distributed in the USA by
Chapman and Hall
New York

Published by International Textbook Company
a member of the Blackie Group
Bishopbriggs, Glasgow G64 2NZ, and
Furnival House, 14–18 High Holborn, London WCIV 6BX

Distributed in the USA by
Chapman and Hall
in association with Methuen, Inc.
733 Third Avenue, New York, N.Y. 10017

British Library Cataloguing in Publication Data

Davis, R.
  Spectral problems in organic chemistry.
  1. Organic compounds—Spectra—Problems,
exercises, etc.
  I. Title II. Wells, C.H.J.
  547.3'0858   QD272.S6

  ISBN 0-7002-0288-9

For the USA, International Standard Book Number is
0-412-00561-1

Printed in Great Britain by Bell and Bain Ltd, Glasgow

Contents

Preface

There have been many advances made in spectroscopy in recent years. One of the most important of these has been the advent of pulsed Fourier Transform n.m.r. spectrometers which enable ^{13}C n.m.r. spectra of organic compounds to be obtained on a routine basis. The increasingly widespread use of such spectrometers has had an impact on the teaching of spectroscopy in relation to structure elucidation, and nowadays as much emphasis is placed on the ^{13}C n.m.r. spectrum of a compound as on its ^{1}H n.m.r. spectrum. It is of importance, therefore, that students be able to interpret the information available in a ^{13}C n.m.r. spectrum. Consequently, we have compiled a set of structural problems in which the ^{13}C n.m.r. spectra of a selection of organic compounds are presented alongside their ^{1}H n.m.r. spectra. In real life, such spectra are used in conjunction with infrared and mass spectra to aid unambiguous structure elucidation. Thus, for each of the problems, the infrared, ^{1}H n.m.r., ^{13}C n.m.r. and mass spectra of the compounds are presented, together with either analytical data or a molecular formula.

The spectral problems are graded in complexity, beginning with those of relative simplicity and ending with what we believe are quite challenging examples. If we have erred towards the side of simplicity in the collection then it is intentional. It is hoped that undergraduate students in their early years at British universities and polytechnics should be able to solve the majority of problems without too much difficulty and so gain experience and confidence in the use of spectral data for the elucidation of organic molecular structures. It is also the intention that the problems should be of value to the teaching of spectral interpretation to undergraduates in other countries.

We should like to thank Mr T. Mills and Mr M. Stephens for assistance in recording the n.m.r. and mass spectra respectively. We should also like to thank Dr A. Cooper for a gift of one of the compounds.

R.D.
C.H.J.W.

Notes on the problems

Analytical data

Analytical data is provided for problems 1–40 inclusive. In all these cases, carbon and hydrogen data are given, together with data for nitrogen and sulphur when these elements are present. In many cases, other elements are present. The molecular formulae of the compounds are given for problems 41 to 56 (except for problem 53).

The number of atoms n of a particular element $X(C, H, N$ or $S)$ in a compound can be obtained from the analysis data using the equation

$$n_X = \frac{(\%X) \times (\text{RMM of compound})}{100 \times (\text{RAM of X})}$$

The relative molecular mass of each compound in the problem set is given by the m/z value of the molecular ion, M^+, in its mass spectrum. In cases where an M^+ ion peak was not observed in the spectrum the m/z value corresponding to the missing M^+ ion peak is given.

Once the number of each of the elements C, H, N or S has been calculated, the other elements present can be deduced from a comparison of the RMM of the compound with the RMM of the entity containing C, H, N or S.

Once the molecular formula of a compound is known, the following equation can be used to calculate the number of units of unsaturation or double bond equivalents, N, in the molecule:

$$N = \tfrac{1}{2}(2n_4 + n_3 - n_1 + 2)$$

where n_4 = number of tetravalent atoms present
n_3 = number of trivalent atoms present
n_1 = number of monovalent atoms present.

Infrared spectra

The infrared spectrum of problem 1 is that obtained from a CCl_4 solution, without solvent subtraction. In all other cases, the spectra presented are those obtained from liquid films or KBr discs. All the spectra were recorded in the transmission mode.

Mass spectra

Mass spectra were recorded at 70 eV. Metastable peaks are shown on the spectra under the listing m*. In all cases, the fragmentation processes associated with these are assigned. On some spectra certain peaks of low intensity are shown with the intensity scale enhanced by a factor of 10.

N.m.r. spectra

^{1}H n.m.r. spectra were recorded on a pulsed instrument operating at 80 MHz. The range 0–10 p.p.m. thus represents 800 Hz. In the cases where signals are observed with chemical shifts greater than 10 p.p.m., the signals are displayed as inserts together with the appropriate chemical shift scale. In some cases, weak multiplets have been recorded at higher sensitivity and these are shown on the spectra. In other cases, complex multiplets are shown on an expanded scale and these are clearly identified. In all cases the spectra were recorded with tetramethylsilane (0 p.p.m.) as internal reference.

The ^{13}C n.m.r. spectra were recorded on the same instrument as the ^{1}H spectra but operating at a frequency of 20.15 MHz. TMS was used as internal reference (0 p.p.m.). All spectra are shown in the proton-noise decoupled form, but chemical shifts and off-resonance multiplicities are tabulated on each spectrum. The multiplicities are denoted in the following way: s, singlet; d, doublet; t, triplet; q, quartet. The ^{13}C spectra of the majority of samples were obtained from $CDCl_3$ solutions. In some of these spectra the absorption due to the solvent can be seen as a triplet at 77.0 p.p.m. Where these peaks occur they are labelled S. The remainder of the spectra were obtained from $(CD_3)_2SO$ and in some of these the solvent signal is observed as a septet centred at 39.7 p.p.m. Again, these peaks are labelled S.

Problems

Problem 1   C. 25.4; H, 3.2%

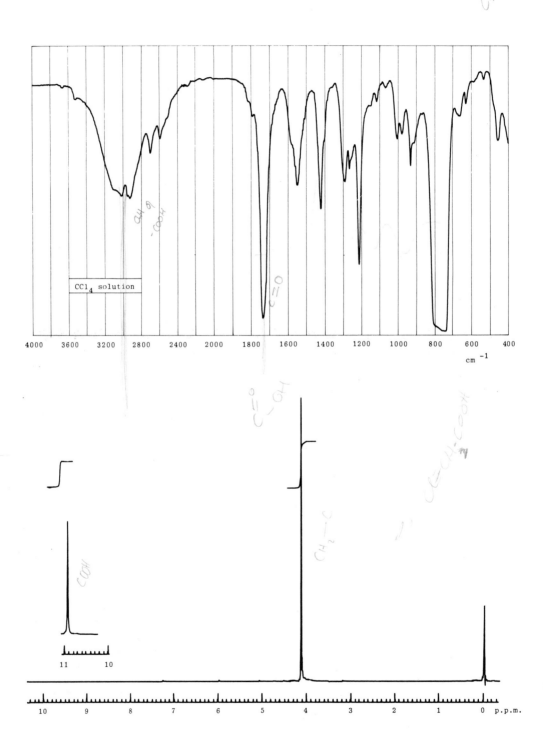

CCl₄ solution

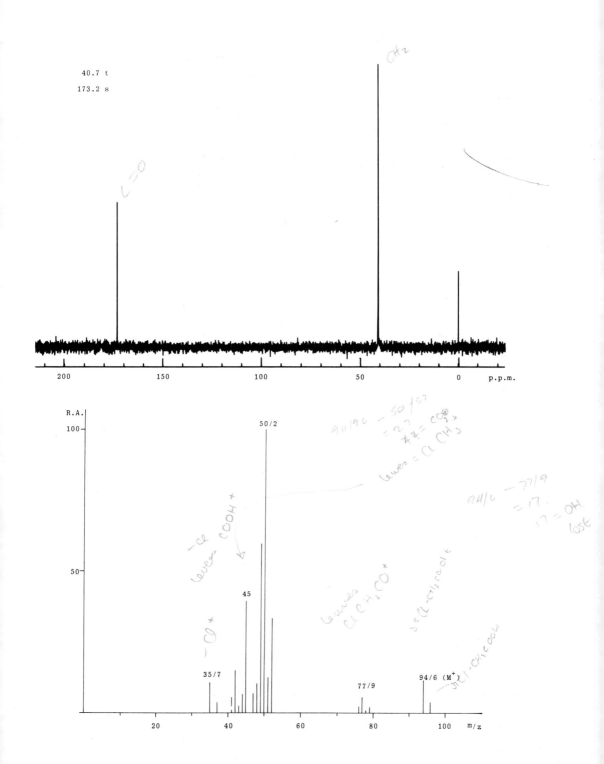

40.7 t

173.2 s

C=O

CH₂

200   150   100   50   0   p.p.m.

R.A.

100 —

50 —

50/2

45

35/7

77/9

94/6 (M⁺)

20   40   60   80   100   m/z

-Cl
lewes COOH⁺

-CO⁺

lewes
Cl CH₂CO⁺

³⁵Cl-CH₂COOH

³⁷Cl-CH₂COOH

94/96 — 50/52
= 27
2 = CO₂⁺
lewes = Cl CH₃

94/6 — 77/9
= 17
17 = OH.
lose

Problem 2   C, 82.1; H, 6.0; N, 12.0%

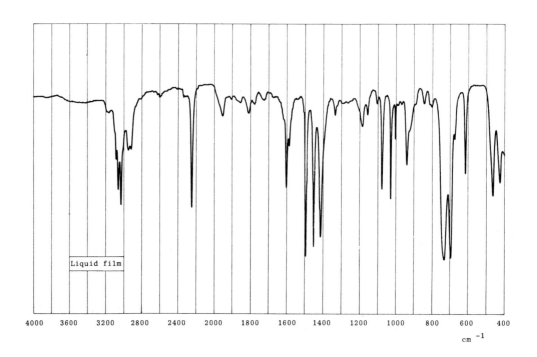

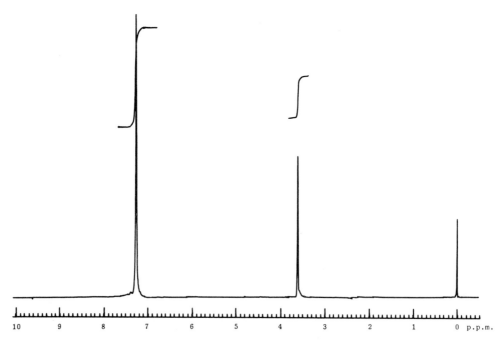

23.1 t
118.4 s
127.8 d
128.0 d
129.0 d
130.6 s

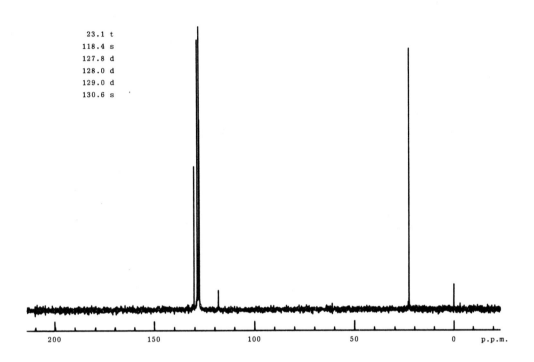

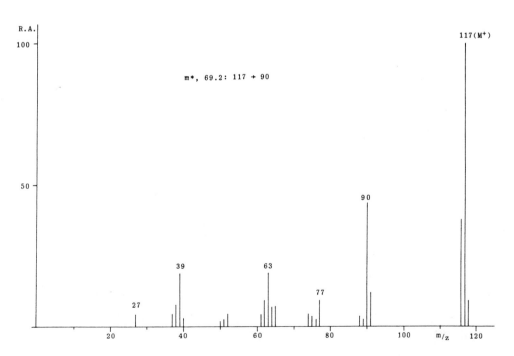

Problem 3   C, 79.3; H, 9.1; N, 11.6%

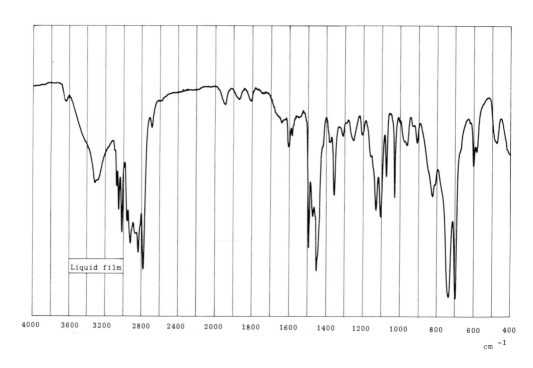

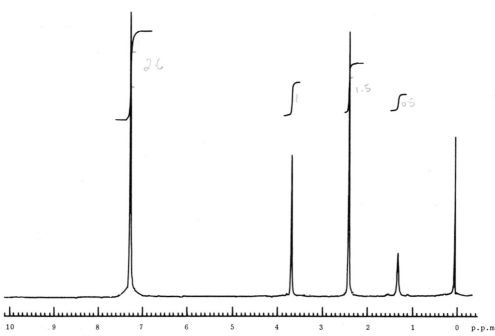

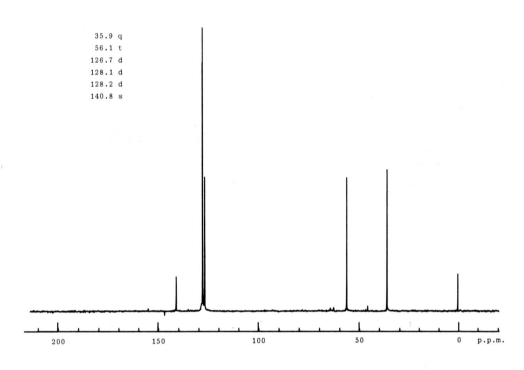

35.9 q
56.1 t
126.7 d
128.1 d
128.2 d
140.8 s

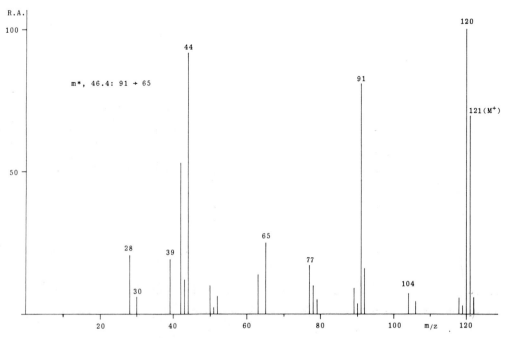

m*, 46.4: 91 → 65

Problem 4   C, 62.1; H, 10.3%

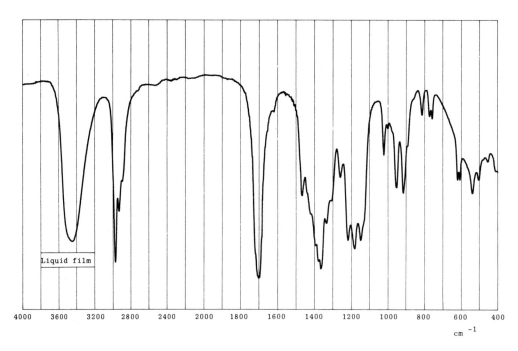

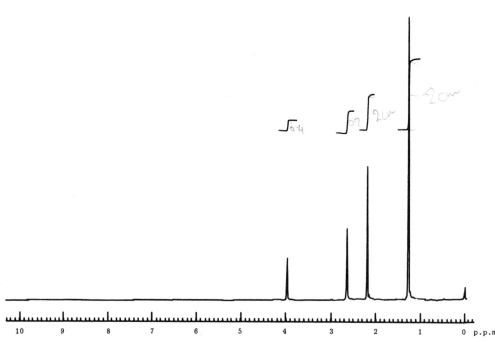

29.4 q
31.8 q
54.7 t
69.6 s
210.0 s

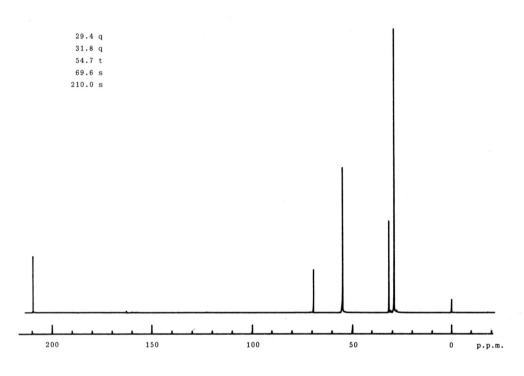

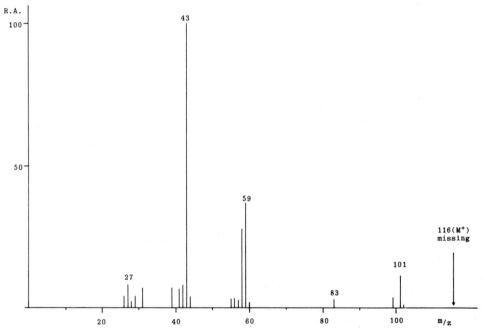

Problem 5   C, 81.8; H, 10.9%

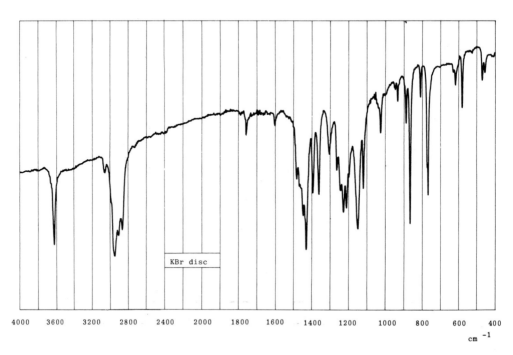

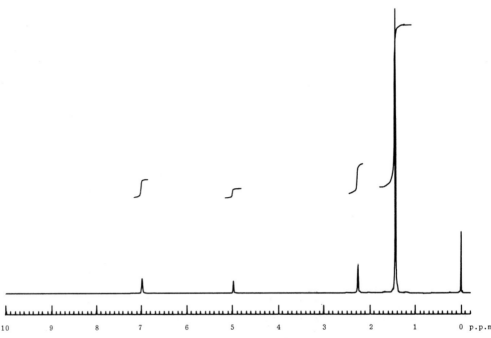

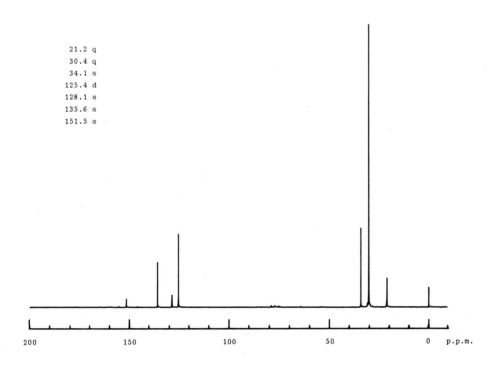

21.2 q
30.4 q
34.1 s
125.4 d
128.1 s
135.6 s
151.5 s

200   150   100   50   0   p.p.m.

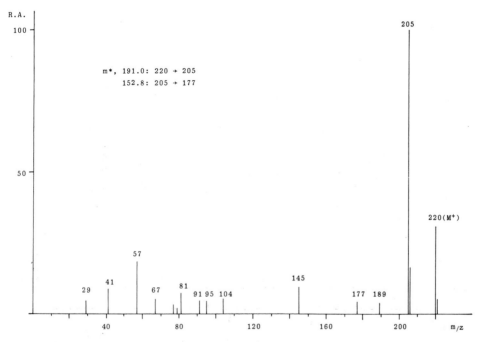

R.A.
100

m*, 191.0: 220 → 205
    152.8: 205 → 177

205

220(M+)

57

29   41   67   81   91 95 104   145   177 189

50

40   80   120   160   200   m/z

B

Problem 6   C, 19.8; H, 0.8; S, 13.2%

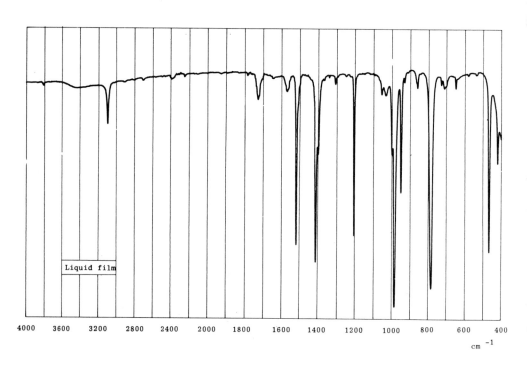

Liquid film

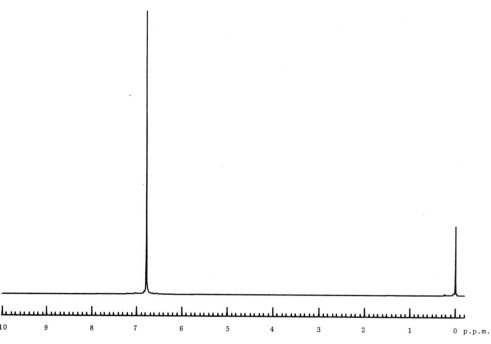

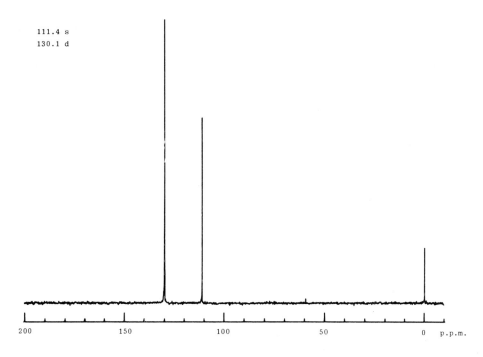

111.4 s
130.1 d

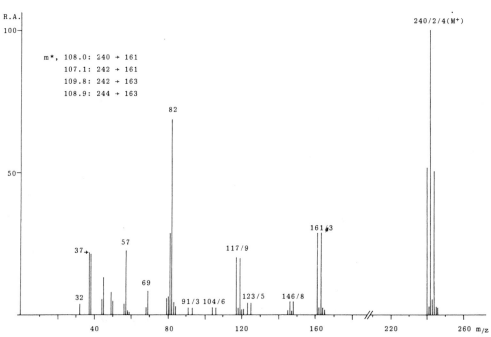

R.A.

240/2/4(M⁺)

m*, 108.0: 240 → 161
107.1: 242 → 161
109.8: 242 → 163
108.9: 244 → 163

82

57

37→

32

69

91/3 104/6

117/9

123/5

146/8

161/3

Problem 7   C, 70.6; H, 13.7%

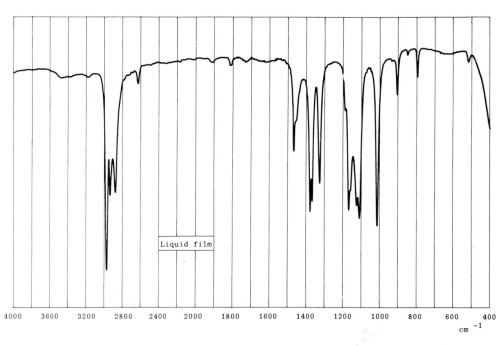

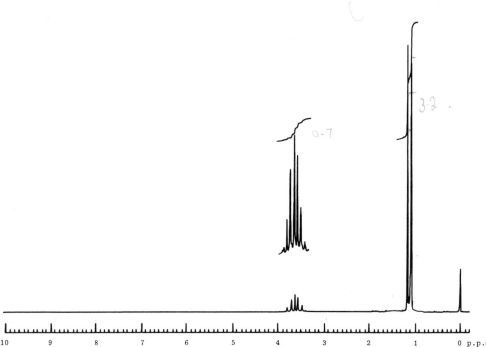

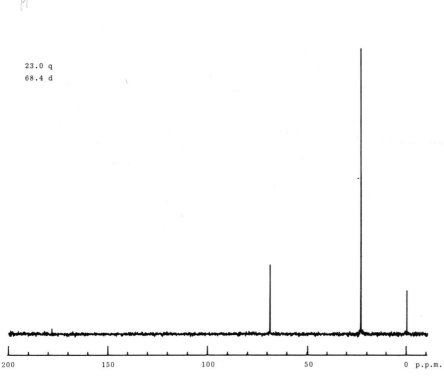

23.0 q
68.4 d

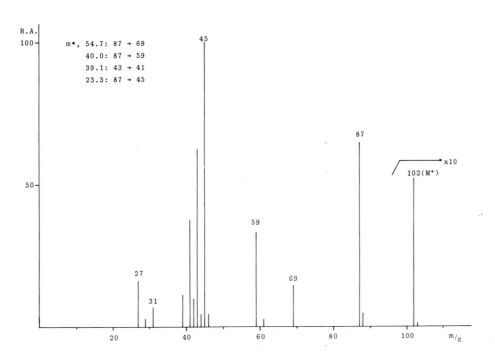

R.A.
100

m*, 54.7: 87 → 69
40.0: 87 → 59
39.1: 43 → 41
23.3: 87 → 45

45

87

x10

102(M⁺)

59

27

31

69

50

20   40   60   80   100   m/z

Problem 8   C, 23.5; H, 3.3%

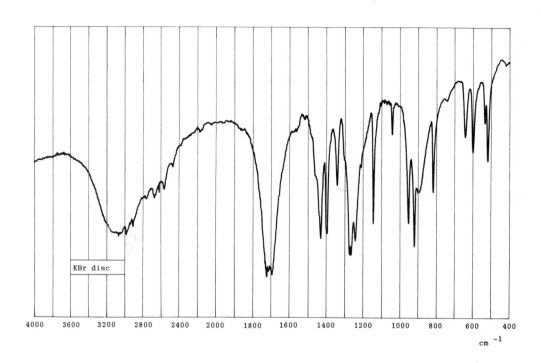

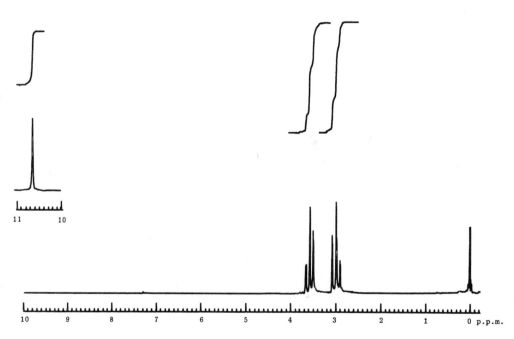

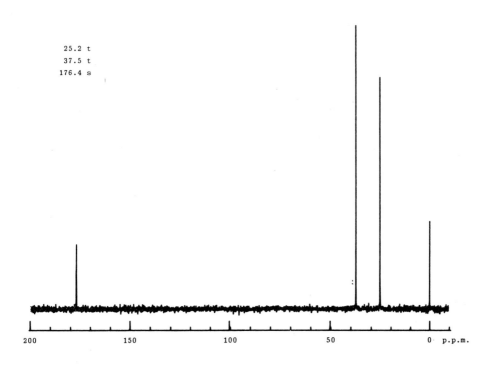

25.2 t
37.5 t
176.4 s

200   150   100   50   0 · p.p.m.

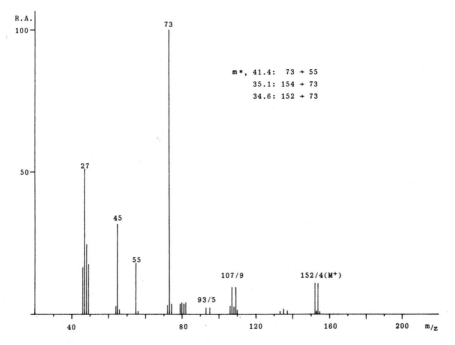

R.A.
100 —

73

m*, 41.4: 73 → 55
35.1: 154 → 73
34.6: 152 → 73

50 —

27

45

55

93/5

107/9

152/4(M⁺)

40   80   120   160   200   m/z

SPECTRAL PROBLEMS IN ORGANIC CHEMISTRY

Problem 9   C, 42.9; H, 2.4; N, 16.7%

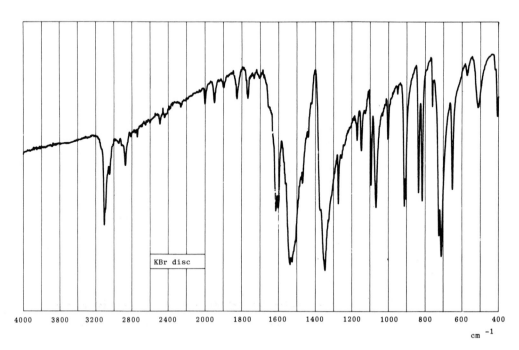

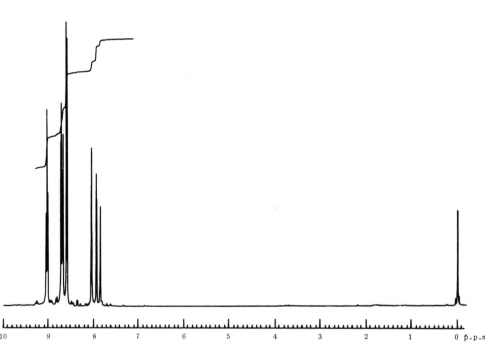

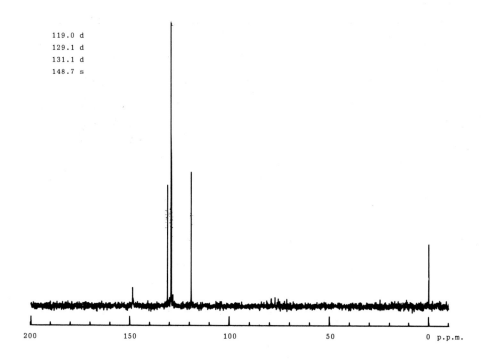

119.0 d
129.1 d
131.1 d
148.7 s

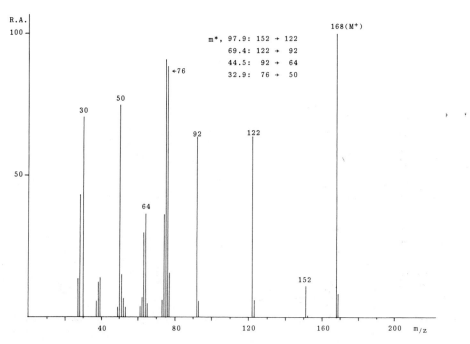

R.A.
100 —

m*, 97.9: 152 → 122
    69.4: 122 → 92
    44.5: 92 → 64
    32.9: 76 → 50

168(M⁺)

Problem 10   C, 54.6; H, 9.1%

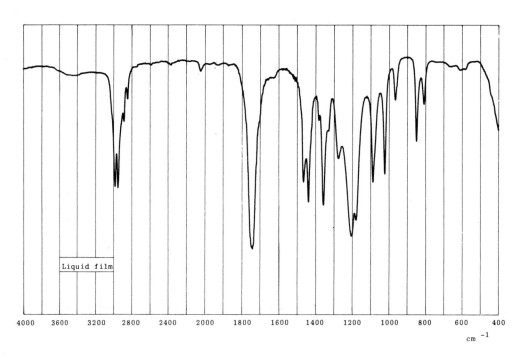

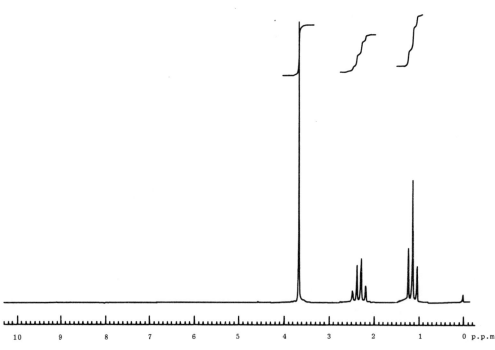

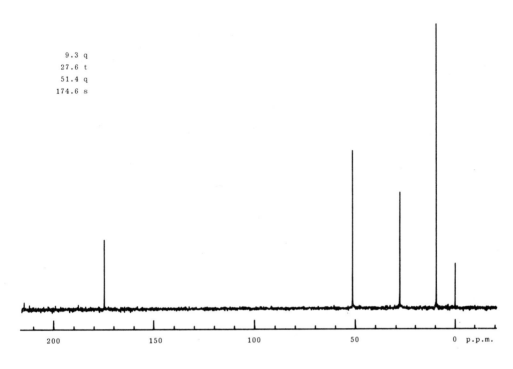

9.3 q
27.6 t
51.4 q
174.6 s

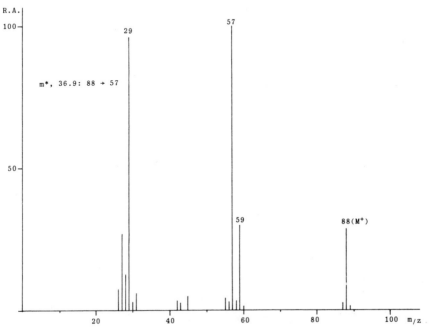

R.A.

100 —

29

57

m*, 36.9: 88 → 57

50 —

59

88(M⁺)

20   40   60   80   100   m/z

Problem 11   C, 53.9; H, 4.5%

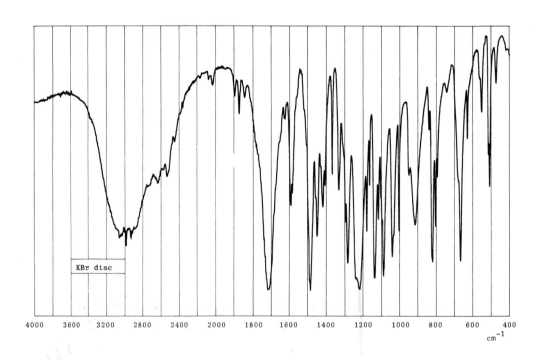

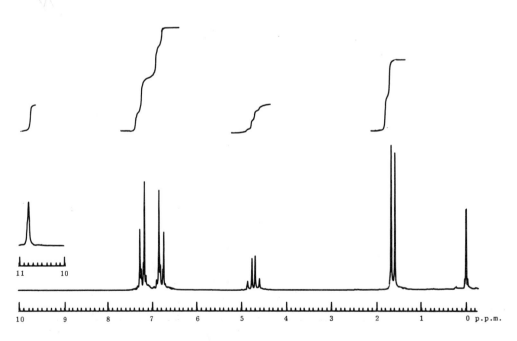

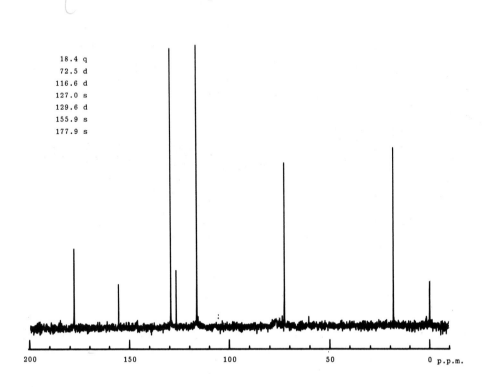

18.4 q
72.5 d
116.6 d
127.0 s
129.6 d
155.9 s
177.9 s

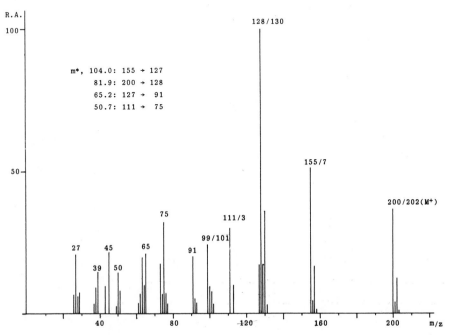

m*, 104.0: 155 → 127
    81.9: 200 → 128
    65.2: 127 → 91
    50.7: 111 → 75

128/130

155/7

200/202(M⁺)

75

111/3

99/101

27

45

65

91

39

50

Problem 12   C, 68.2; H, 13.6%

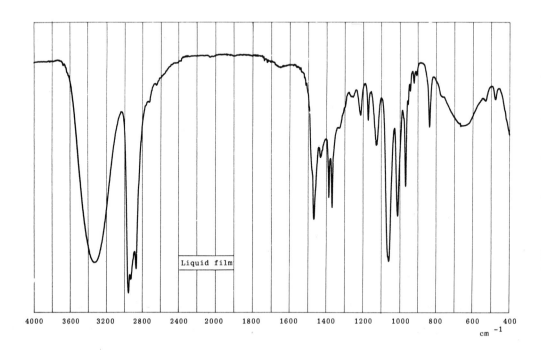

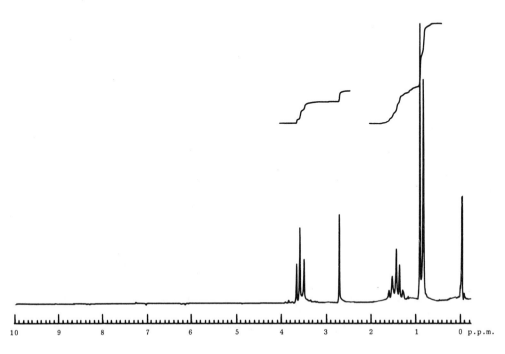

22.7 q
25.0 d
41.8 t
60.5 t

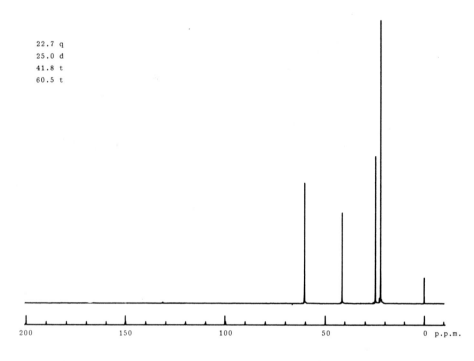

200   150   100   50   0 p.p.m.

R.A.

100—

55

42

70

m*, 43.2: 70 → 55
    39.1: 43 + 41

50—

31

88(M⁺)
missing

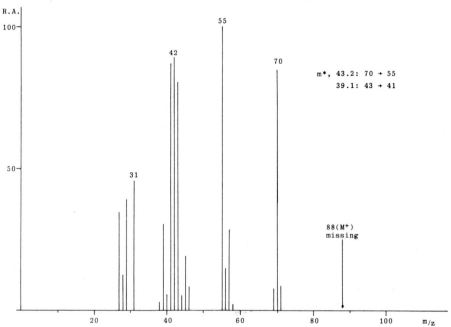

20   40   60   80   100   m/z

Problem 13   C, 68.2; H, 13.6%

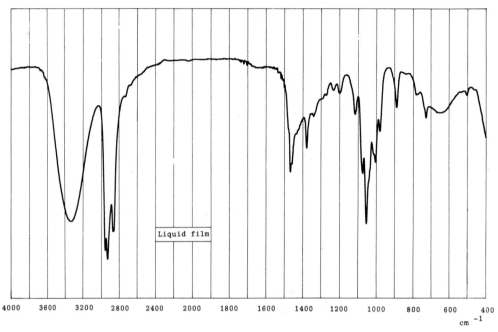

Liquid film

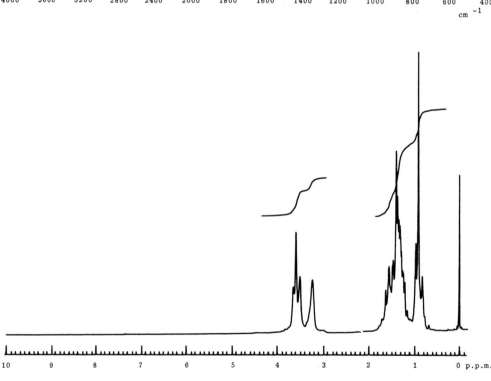

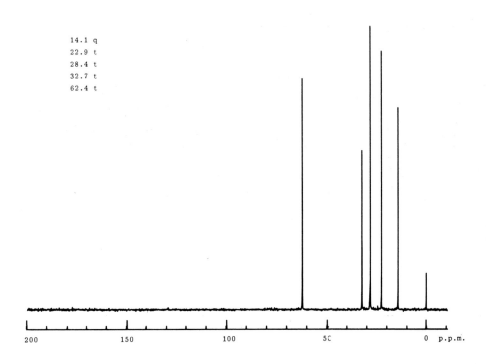

14.1 q
22.9 t
28.4 t
32.7 t
62.4 t

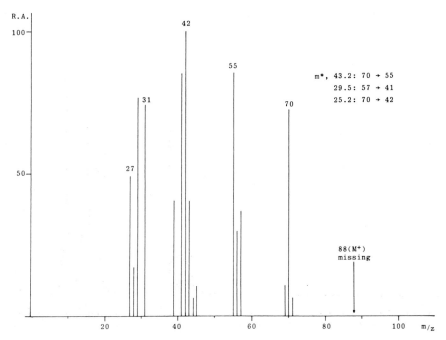

m*, 43.2: 70 → 55
29.5: 57 → 41
25.2: 70 → 42

88(M⁺)
missing

C

Problem 14   C, 30.5; H, 1.7%

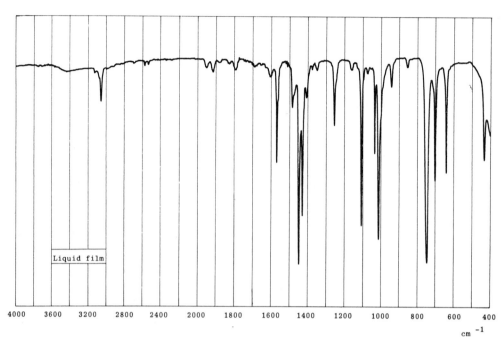

Liquid film

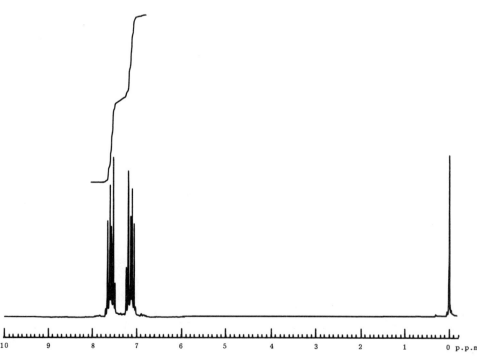

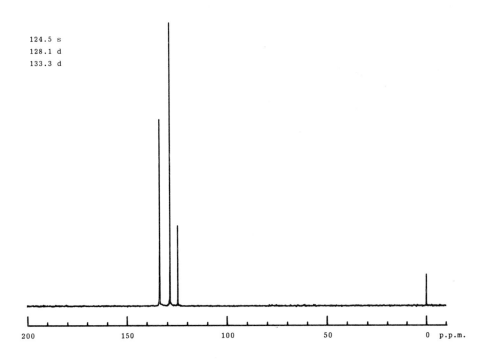

124.5 s
128.1 d
133.3 d

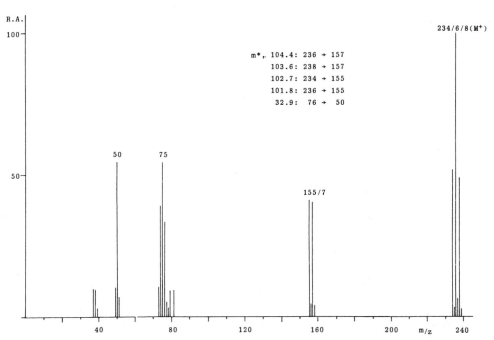

R.A.

100 —

234/6/8(M$^+$)

m*, 104.4: 236 → 157
103.6: 238 → 157
102.7: 234 → 155
101.8: 236 → 155
32.9: 76 → 50

50

75

50 —

155/7

Problem 15   C, 84.2; H, 15.8%

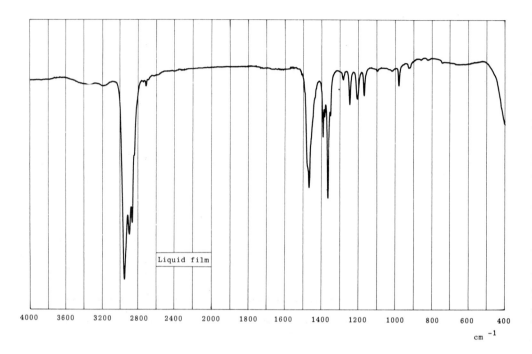

Liquid film

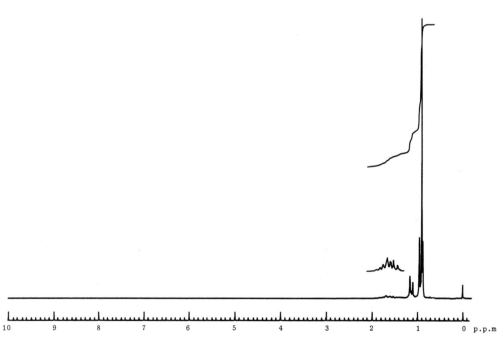

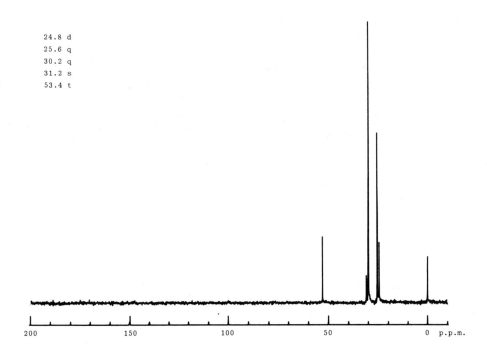

24.8 d
25.6 q
30.2 q
31.2 s
53.4 t

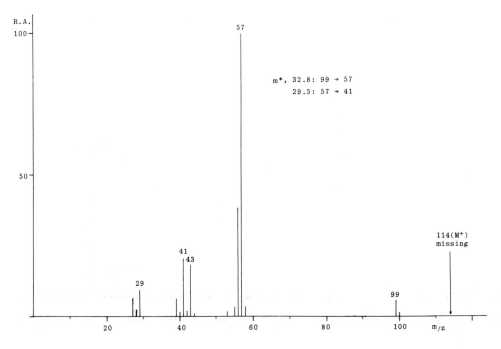

R.A.
100 —

57

m*, 32.8: 99 → 57
      29.5: 57 → 41

114(M⁺)
missing

50 —

41
43

29

99

20   40   60   80   100   m/z

Problem 16   C, 71.6; H, 7.5; N, 20.9%

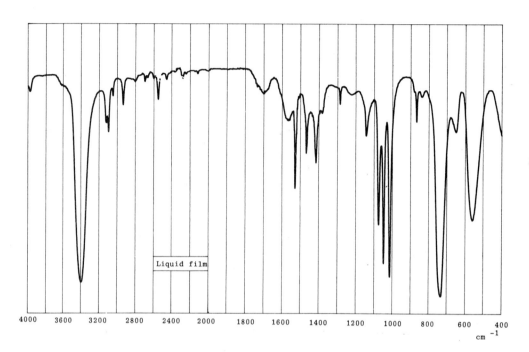

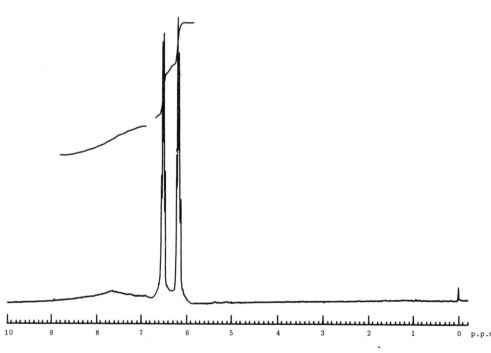

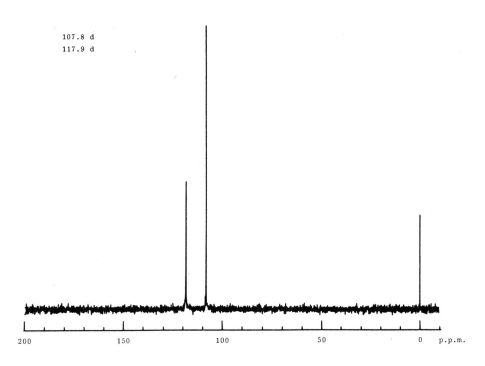

107.8 d
117.9 d

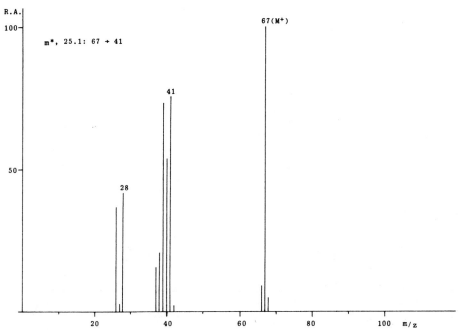

R.A.

100—

m*, 25.1: 67 → 41

67(M⁺)

41

28

50—

20   40   60   80   100   m/z

Problem 17   C, 88.9; H, 11.1%

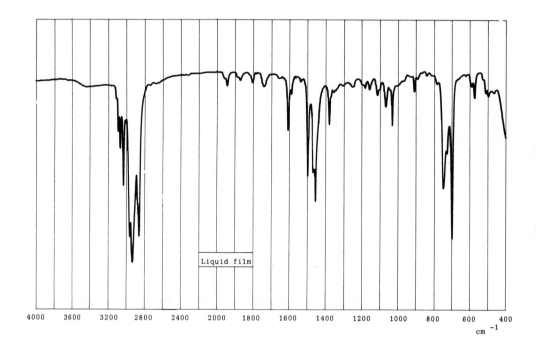

Liquid film

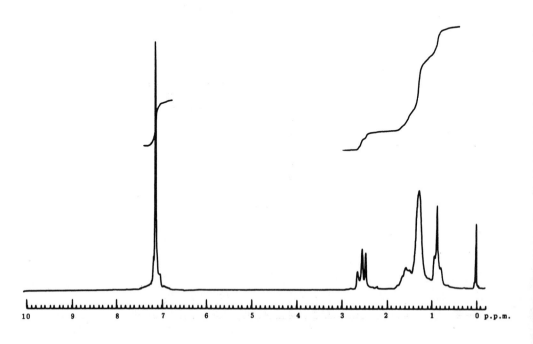

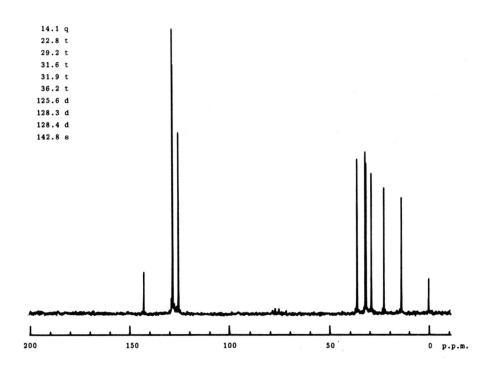

14.1 q
22.8 t
29.2 t
31.6 t
31.9 t
36.2 t
125.6 d
128.3 d
128.4 d
142.8 s

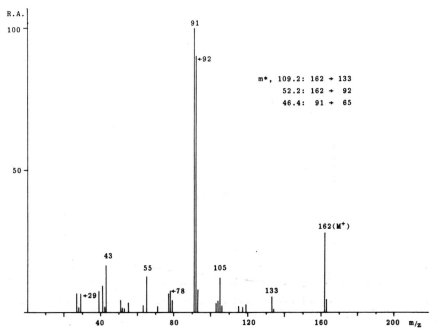

R.A.

m*, 109.2: 162 → 133
52.2: 162 → 92
46.4: 91 → 65

Problem 18   C, 73.5; H, 10.2%

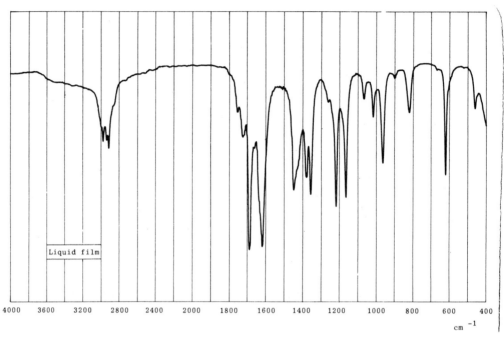

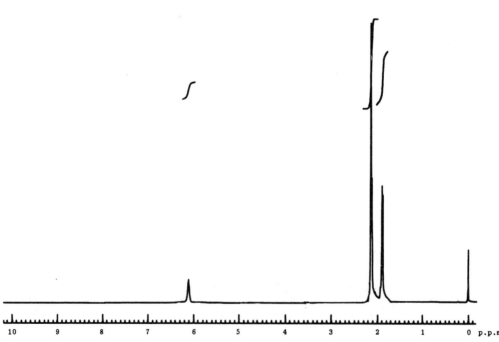

20.4 q
27.4 q
31.4 q
124.4 d
154.6 s
197.9 s

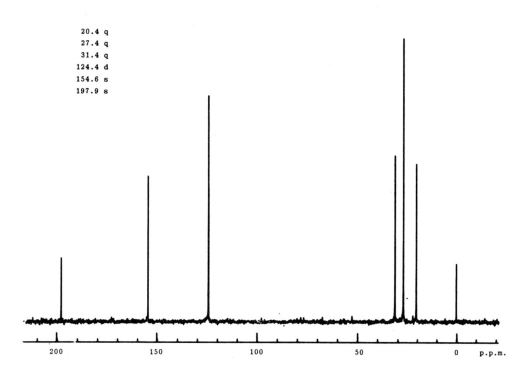

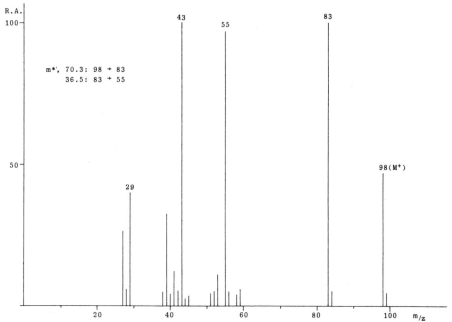

Problem 19   C, 67.0; H, 7.3; N, 7.8%

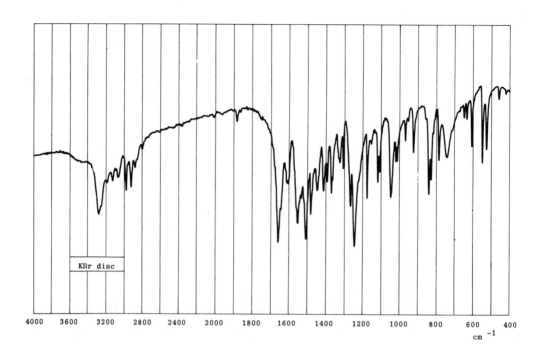

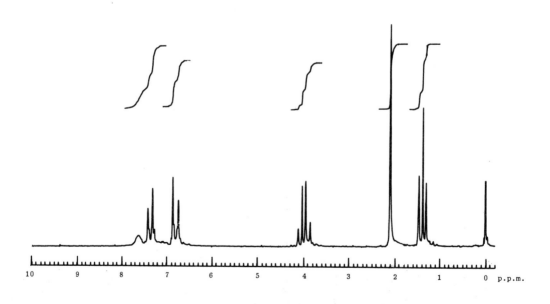

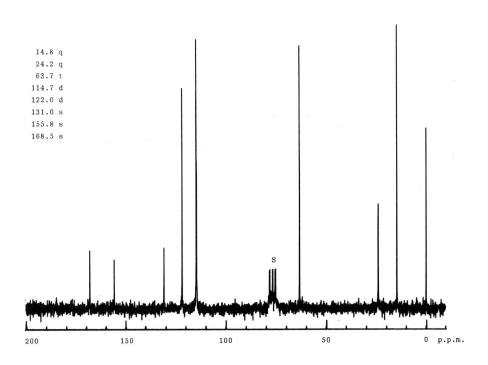

14.8 q
24.2 q
63.7 t
114.7 d
122.0 d
131.0 s
155.8 s
168.5 s

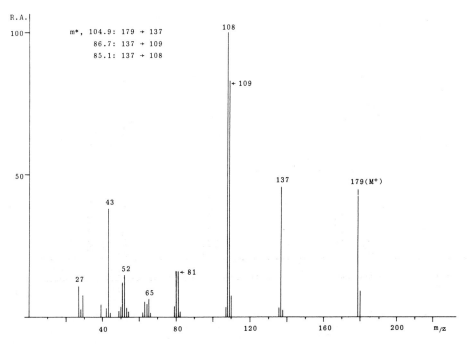

R.A.

100 —

m*, 104.9: 179 → 137
86.7: 137 → 109
85.1: 137 → 108

108

← 109

50 —

43

137

179(M⁺)

27

52

← 81

65

40   80   120   160   200   m/z

Problem 20   C, 60.0; H, 8.0%

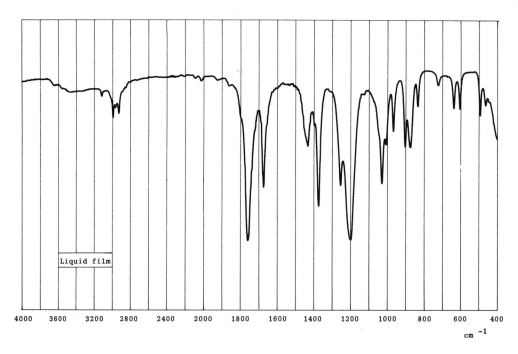

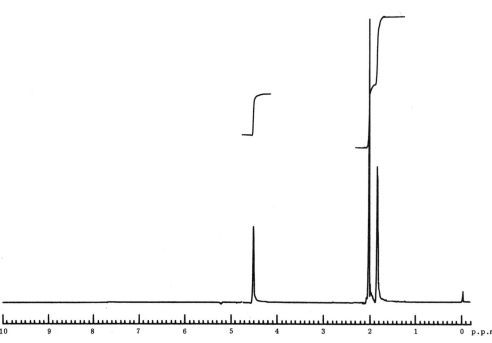

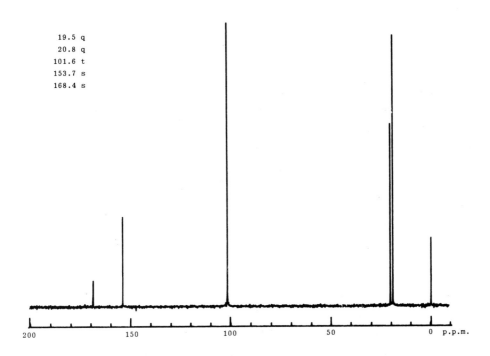

```
19.5 q
20.8 q
101.6 t
153.7 s
168.4 s
```

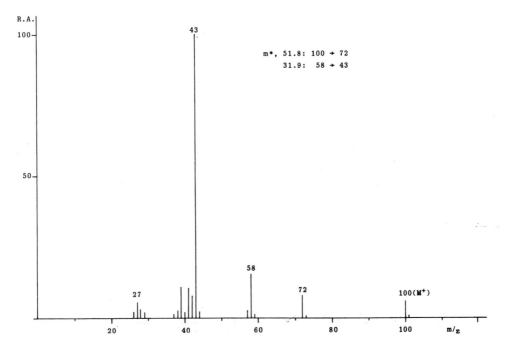

m*, 51.8: 100 → 72
31.9: 58 → 43

Problem 21   C, 80.0; H, 6.7%

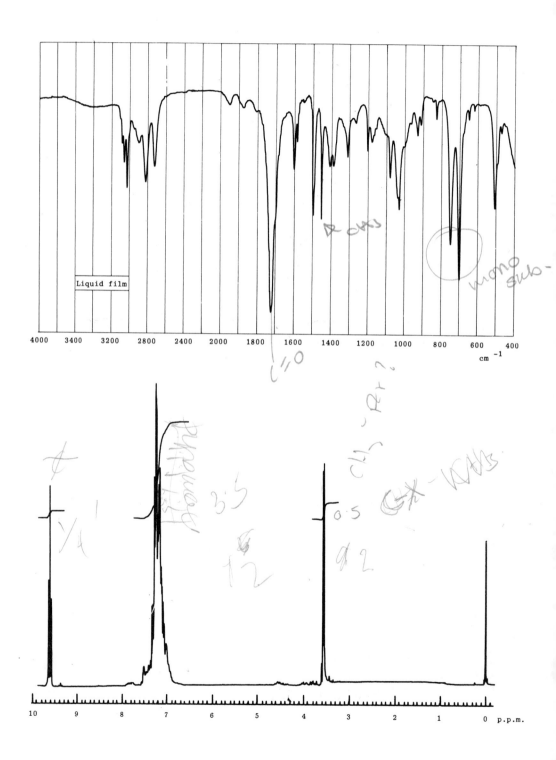

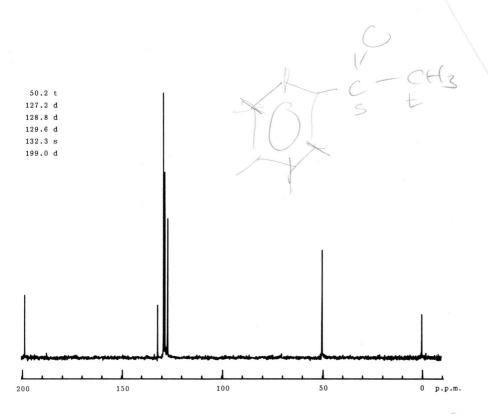

```
 50.2 t
127.2 d
128.8 d
129.6 d
132.3 s
199.0 d
```

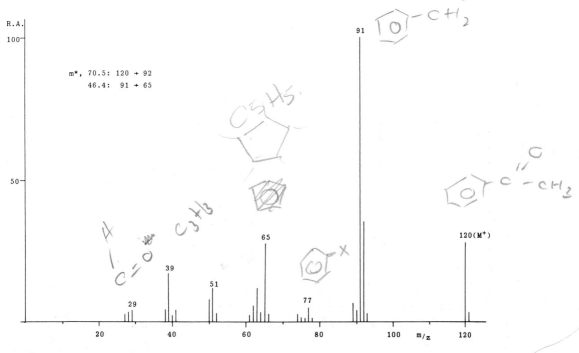

R.A.

100

m*, 70.5: 120 → 92
    46.4: 91 → 65

91

120(M⁺)

65

39

51

77

29

D

Problem 22   C, 55.8; H, 7.0%

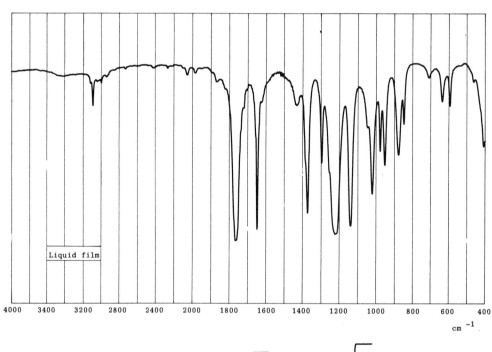

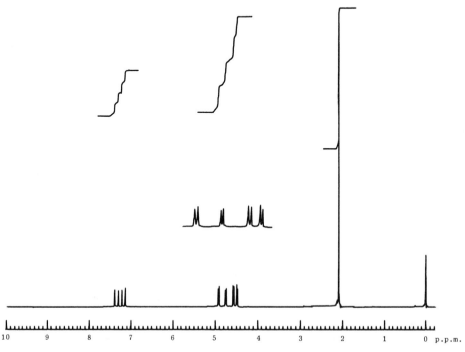

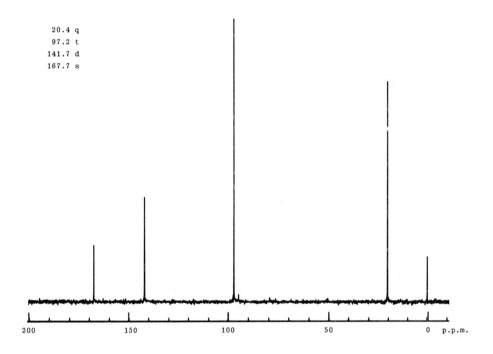

20.4 q
97.2 t
141.7 d
167.7 s

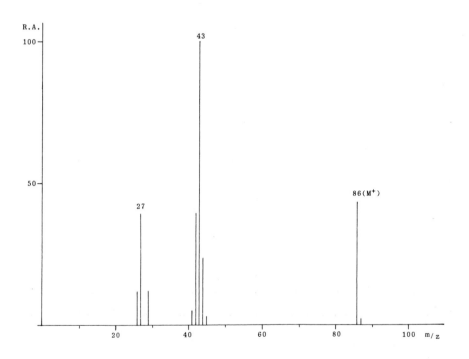

Problem 23   C, 65.5; H, 6.7; N, 8.5%

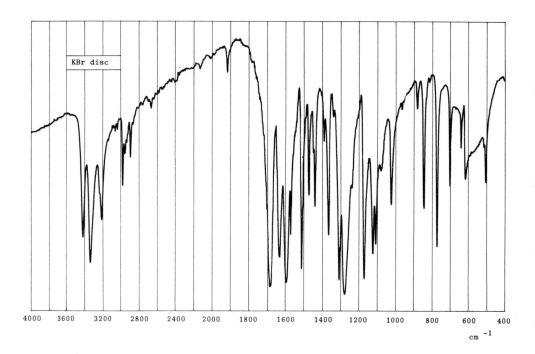

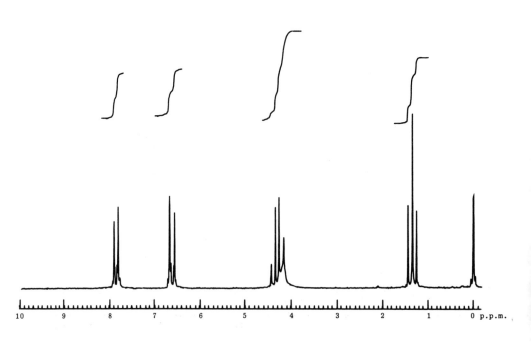

14.4 q
60.3 t
113.7 d
119.5 s
131.5 d
151.4 s
166.9 s

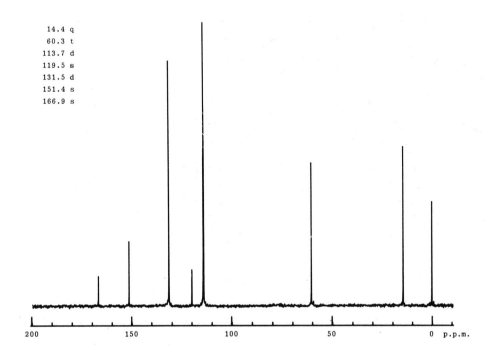

200   150   100   50   0 p.p.m.

R.A.
100 –

m*, 113.7: 165 → 137
      87.3: 165 → 120
      70.5: 120 → 92
      46.0: 92 → 65

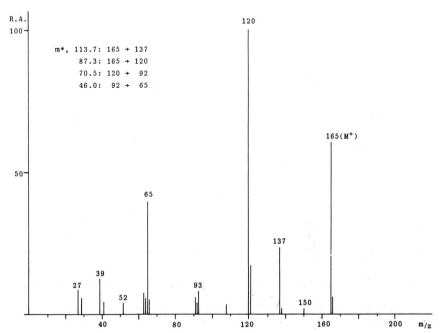

50 –

120

165(M⁺)

65

137

27   39

52   93   150

27   39   52   93

40   80   120   160   200   m/z

Problem 24   C, 29.8; H, 4.1%

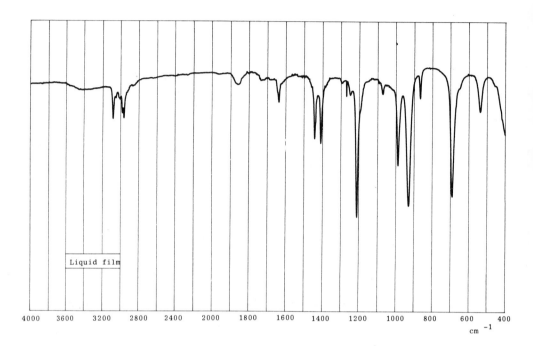

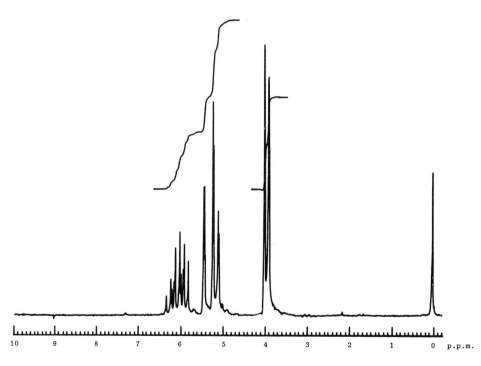

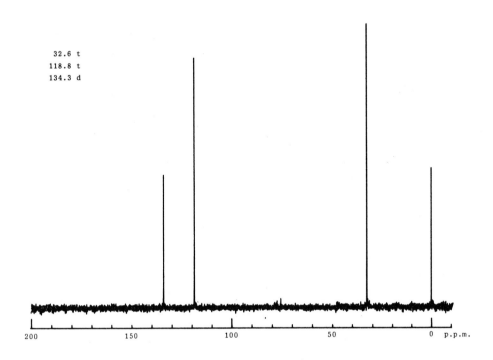

32.6 t
118.8 t
134.3 d

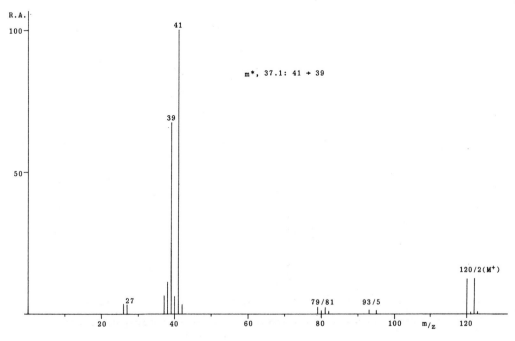

Problem 25 C, 72.0; H, 12.0%

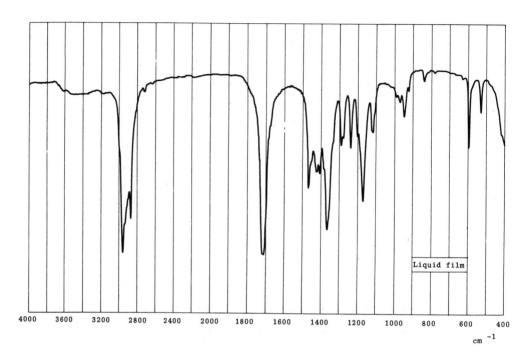

Liquid film

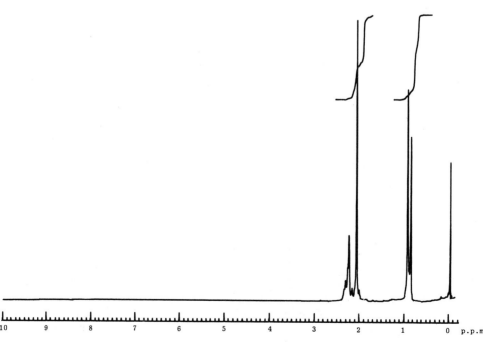

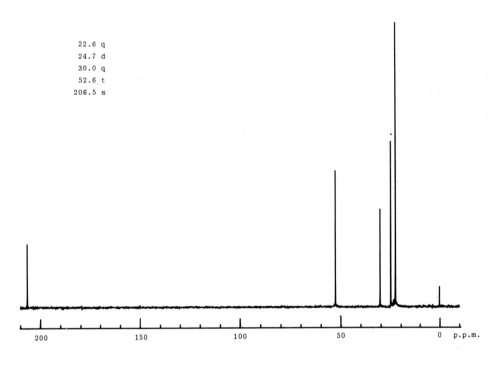

22.6 q
24.7 d
30.0 q
52.6 t
206.5 s

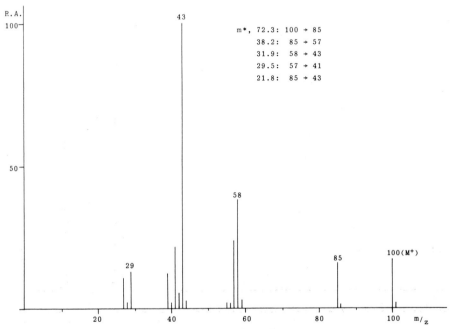

m*, 72.3: 100 → 85
38.2: 85 → 57
31.9: 58 → 43
29.5: 57 → 41
21.8: 85 → 43

Problem 26   C, 73.0; H, 5.4%

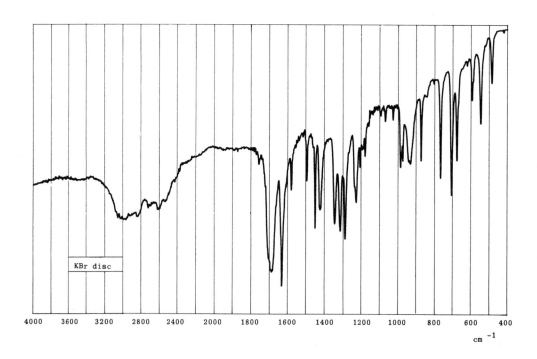

cm $^{-1}$

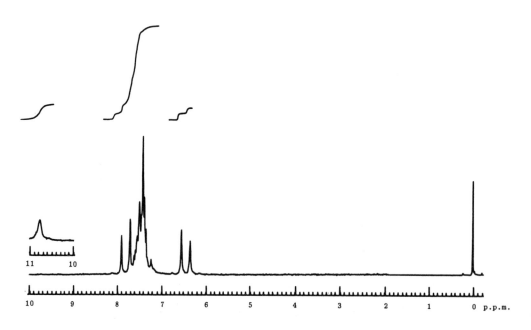

117.5 d
128.3 d
128.9 d
130.7 s
134.1 d
147.0 d
172.6 s

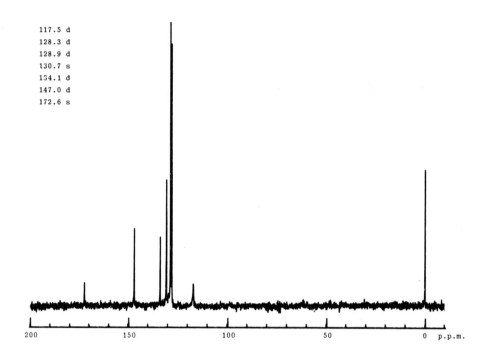

200   150   100   50   0   p.p.m.

R.A.

100 —

m*, 146.0: 148 → 147
81.0: 131 → 103
72.2: 147 → 103
57.6: 103 → 77

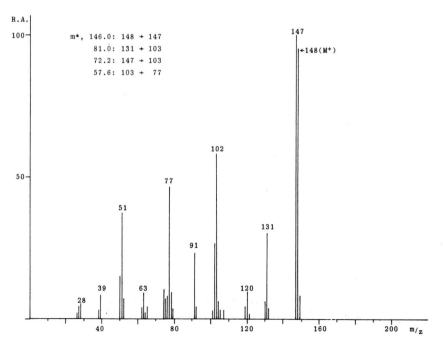

147
←148(M⁺)

102

77

51

91

131

50 —

28   39   63   120

40   80   120   160   200   m/z

Problem 27   C, 73.5; H, 10.2%

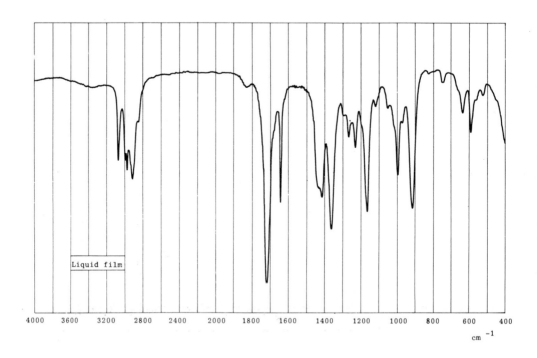

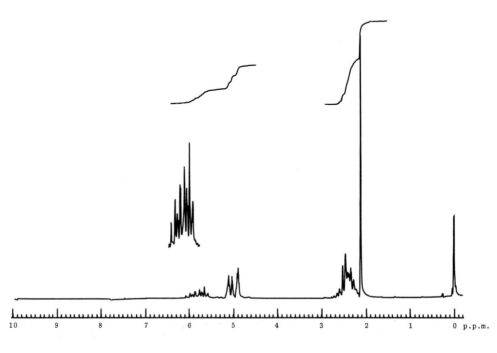

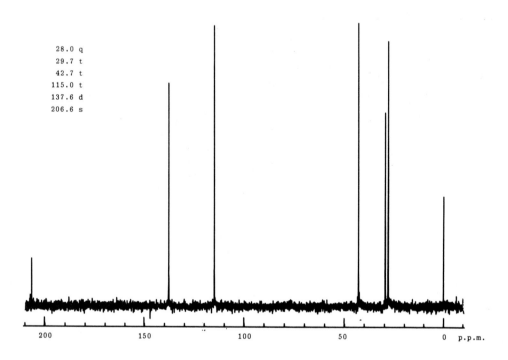

28.0 q
29.7 t
42.7 t
115.0 t
137.6 d
206.6 s

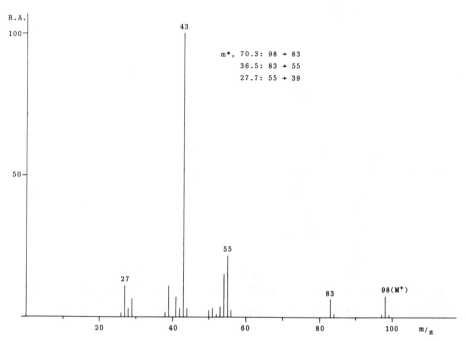

m*, 70.3: 98 → 83
36.5: 83 → 55
27.7: 55 → 39

Problem 28   C, 60.3; H, 5.0; N, 7.8%

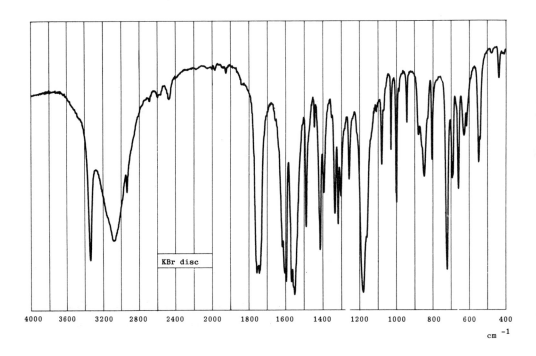

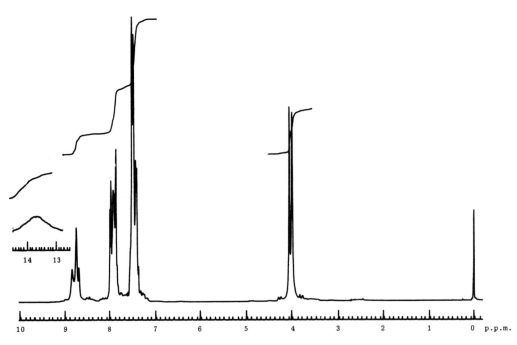

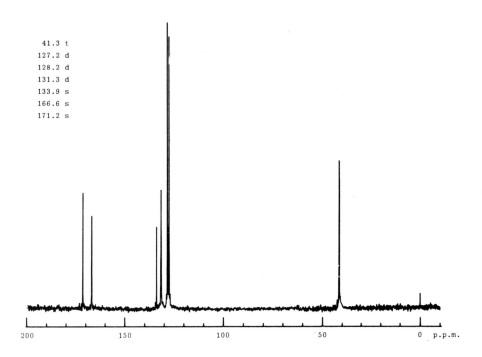

```
41.3  t
127.2 d
128.2 d
131.3 d
133.9 s
166.6 s
171.2 s
```

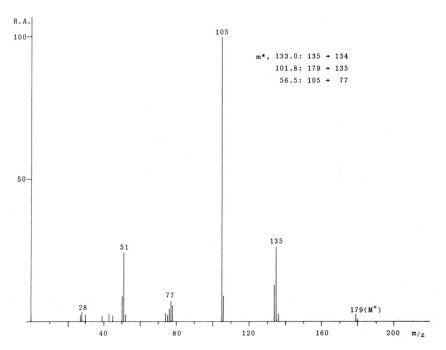

```
m*, 133.0: 135 → 134
    101.8: 179 → 135
     56.5: 105 →  77
```

Problem 29   C, 73.7; H, 12.3%

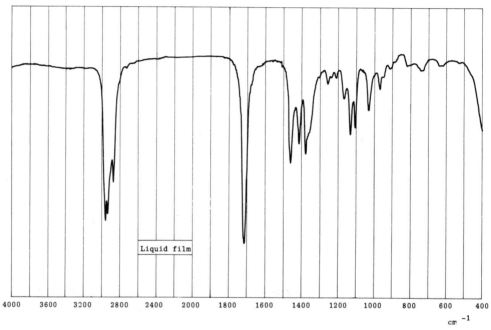

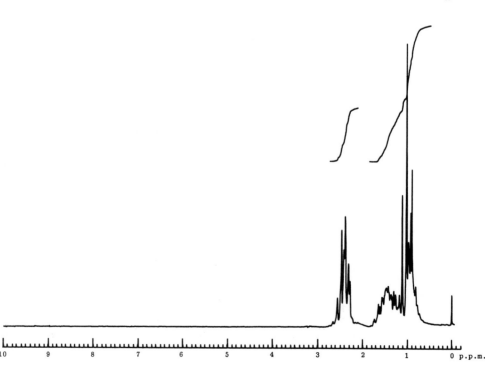

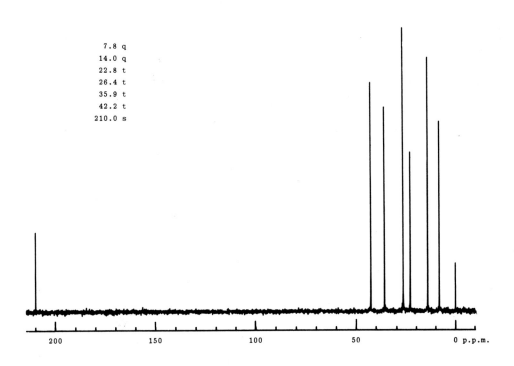

```
7.8 q
14.0 q
22.8 t
26.4 t
35.9 t
42.2 t
210.0 s
```

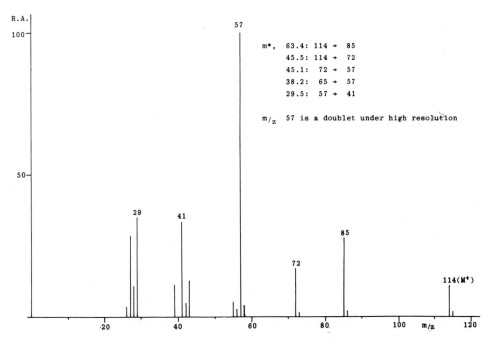

```
m*,  63.4: 114 → 85
     45.5: 114 → 72
     45.1:  72 → 57
     38.2:  65 → 57
     29.5:  57 → 41

m/z  57 is a doublet under high resolution
```

E

Problem 30   C, 55.2; H, 10.3; N, 16.1%

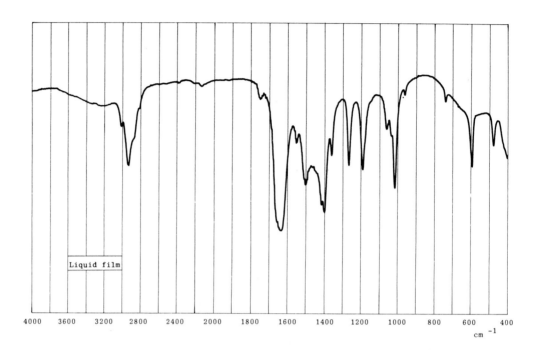

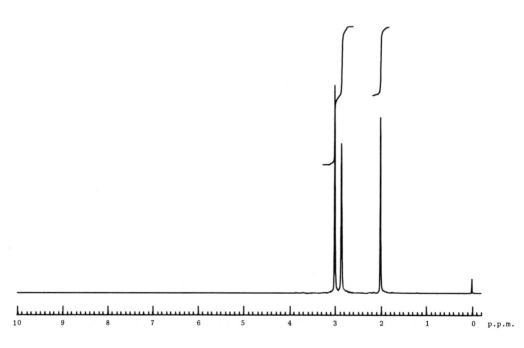

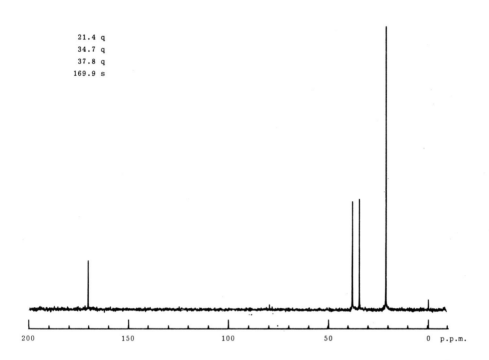

```
21.4 q
34.7 q
37.8 q
169.9 s
```

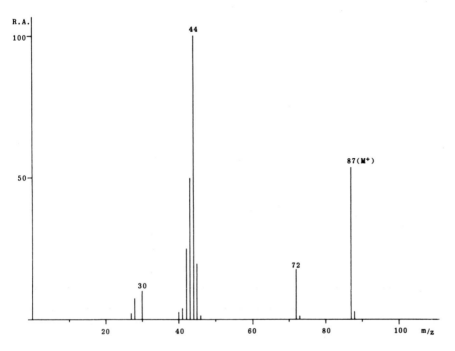

Problem 31   C, 73.5; H, 10.2%

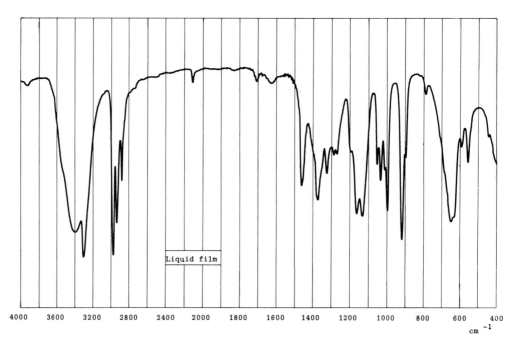

Liquid film

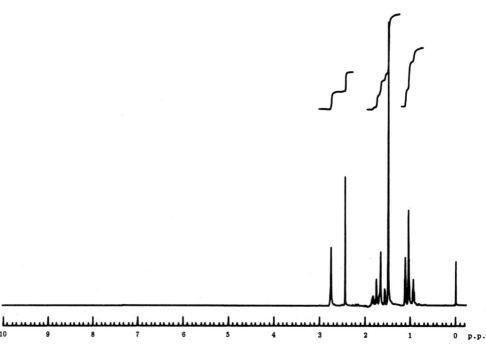

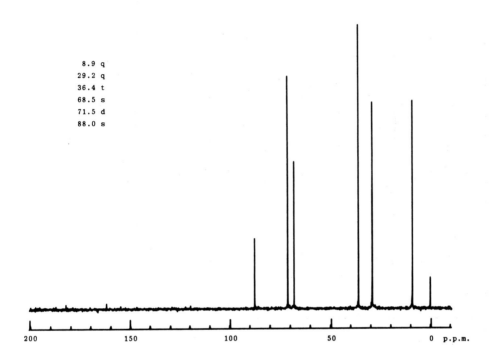

8.9 q
29.2 q
36.4 t
68.5 s
71.5 d
88.0 s

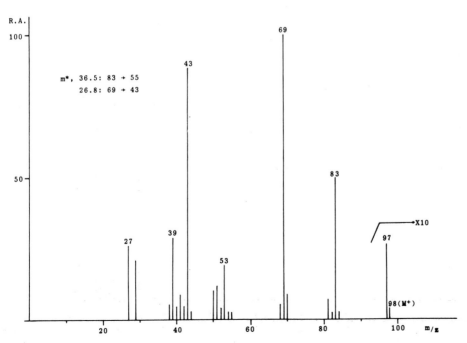

m*, 36.5: 83 → 55
26.8: 69 → 43

Problem 32   C, 61.2; H, 6.1%

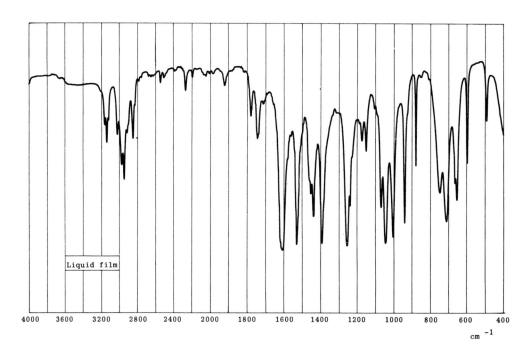

Liquid film

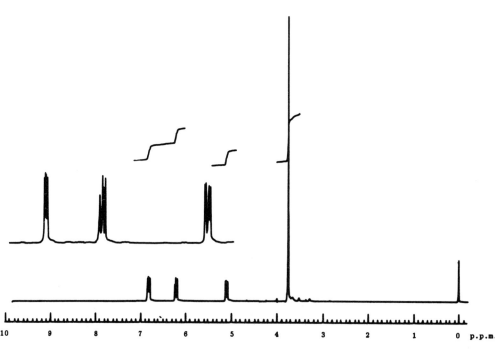

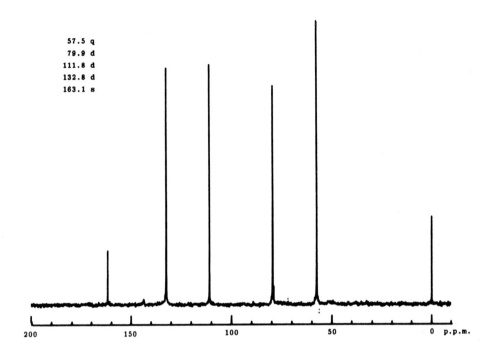

57.5 q
79.9 d
111.8 d
132.8 d
163.1 s

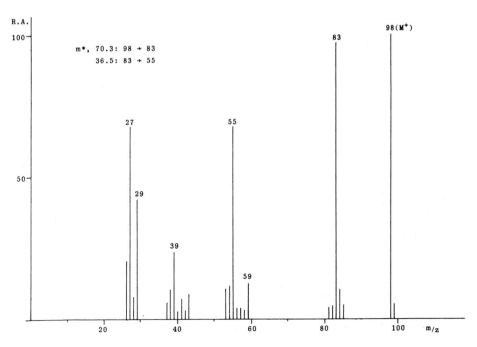

m*, 70.3: 98 → 83
36.5: 83 → 55

Problem 33   C, 28.2; H, 1.6; S, 12.5%

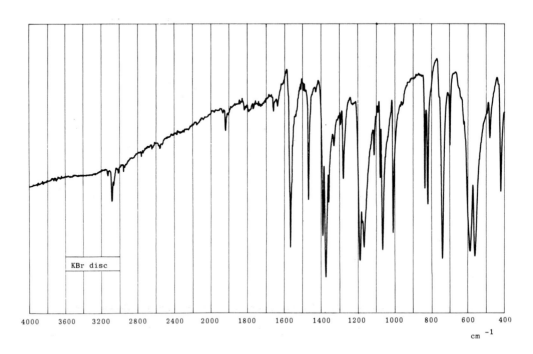

KBr disc

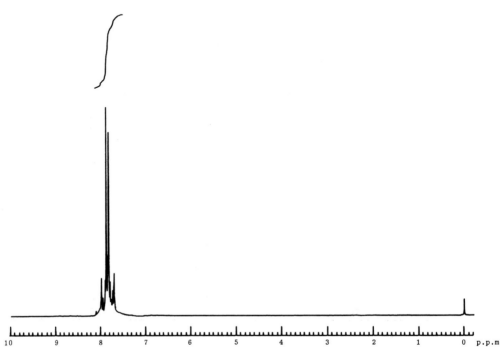

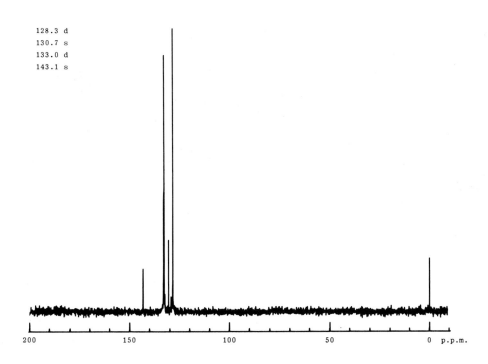

128.3 d
130.7 s
133.0 d
143.1 s

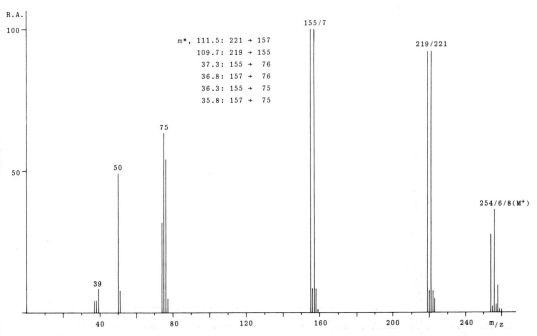

m*, 111.5: 221 → 157
109.7: 219 → 155
37.3: 155 → 76
36.8: 157 → 76
36.3: 155 → 75
35.8: 157 → 75

Problem 34   C, 63.2; H, 8.8%

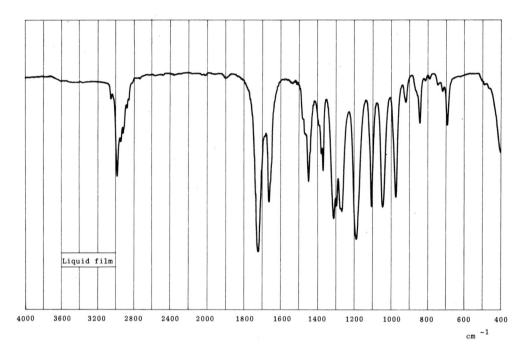

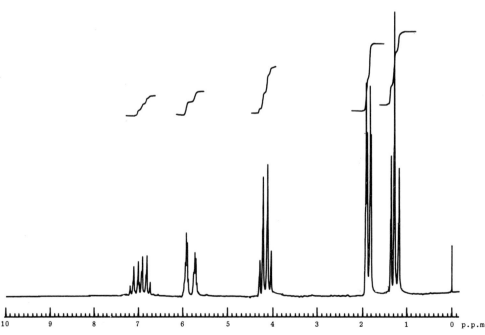

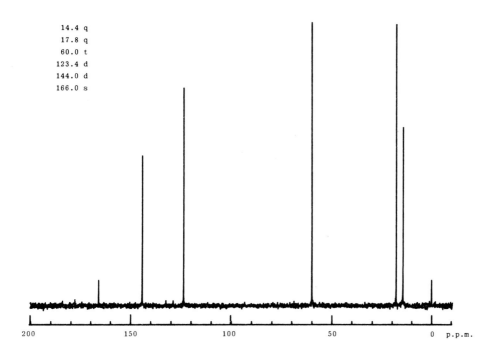

14.4 q
17.8 q
60.0 t
123.4 d
144.0 d
166.0 s

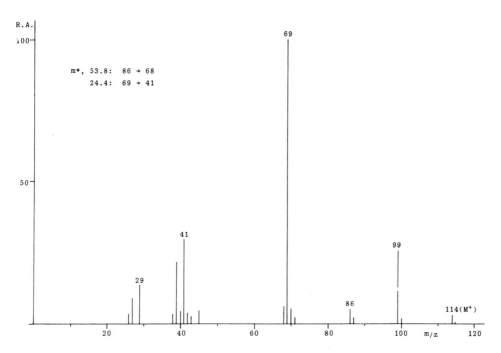

m*, 53.8: 86 → 68
24.4: 69 → 41

Problem 35   C, 39.1; H, 2.2; N, 15.2%

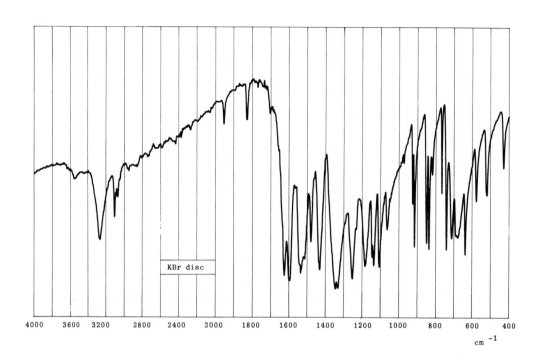

KBr disc

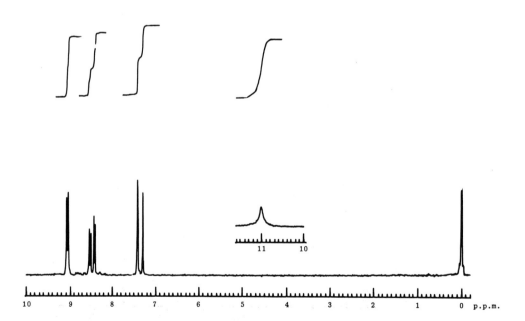

121.3 d
121.9 d
131.6 d
132.1 s
140.1 s
159.1 s

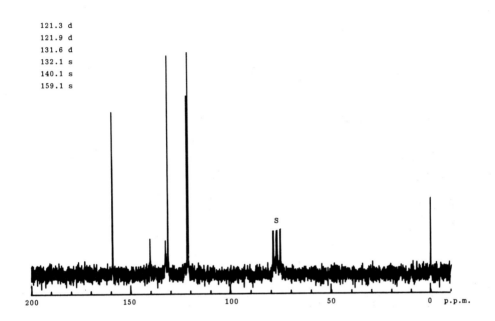

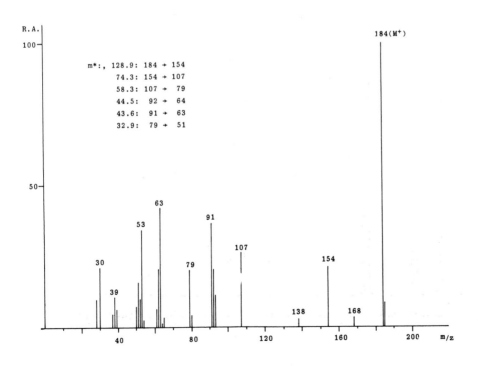

Problem 36   C, 68.9; H, 4.9%

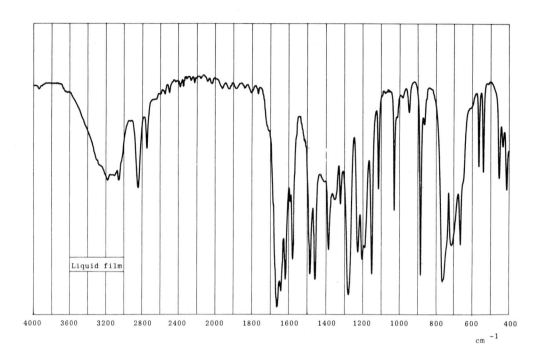

Liquid film

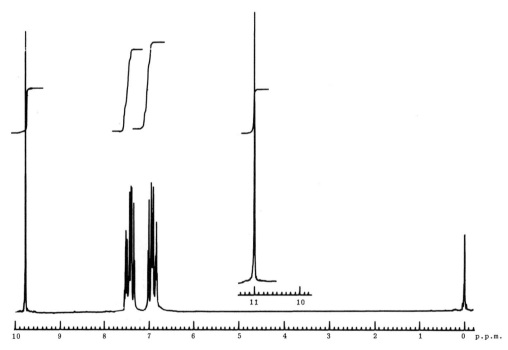

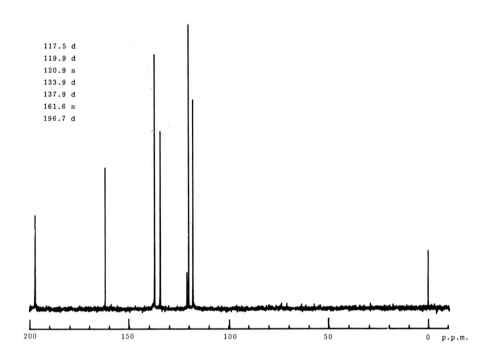

117.5 d
119.9 d
120.9 s
133.9 d
137.9 d
161.6 s
196.7 d

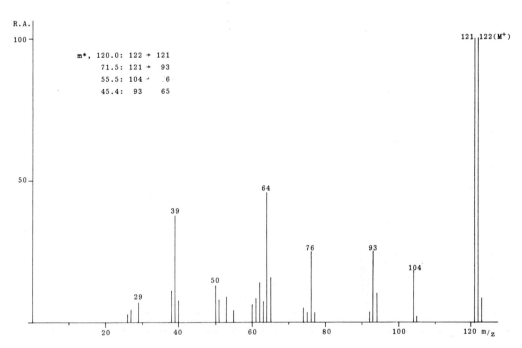

m*, 120.0: 122 → 121
     71.5: 121 → 93
     55.5: 104 → 6
     45.4: 93   65

121 122(M⁺)

Problem 37   C, 77.8; H, 7.4%

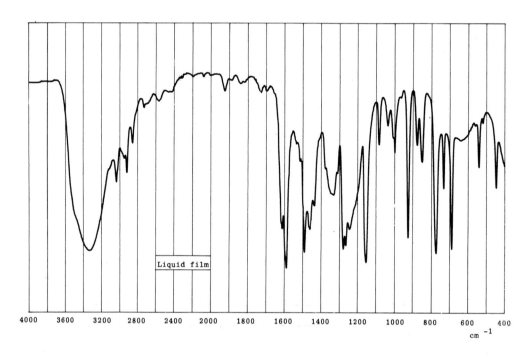

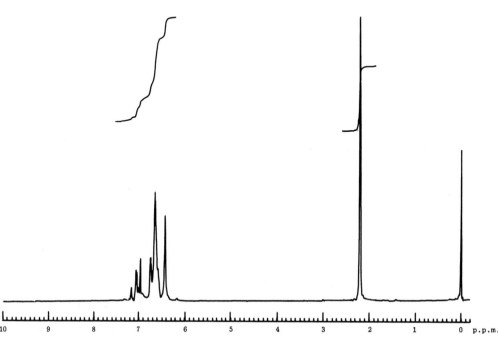

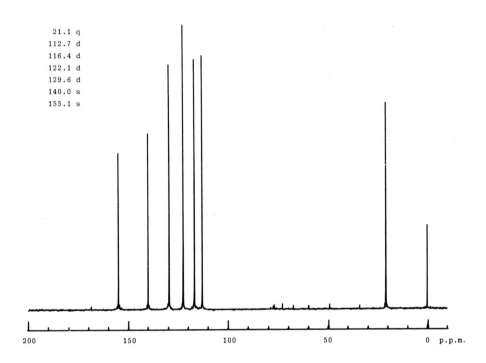

21.1 q
112.7 d
116.4 d
122.1 d
129.6 d
140.0 s
155.1 s

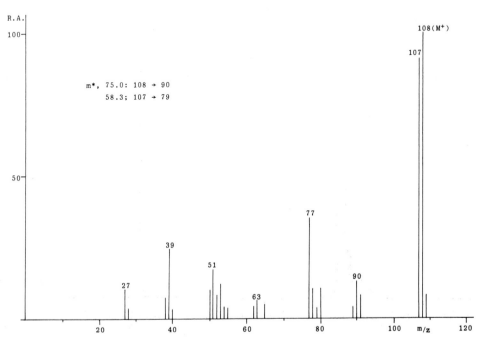

R.A.

m*, 75.0: 108 → 90
     58.3; 107 → 79

108(M⁺)
107
77
39
51
27
63
90

F

Problem 38   C, 80.0; H, 6.7%

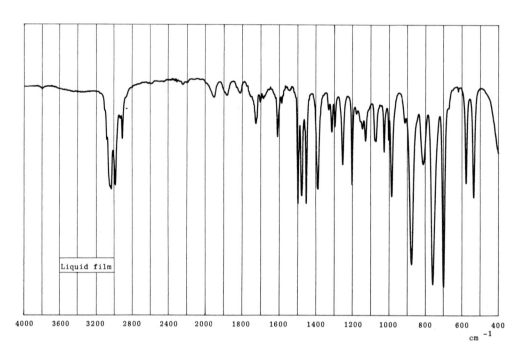

Liquid film

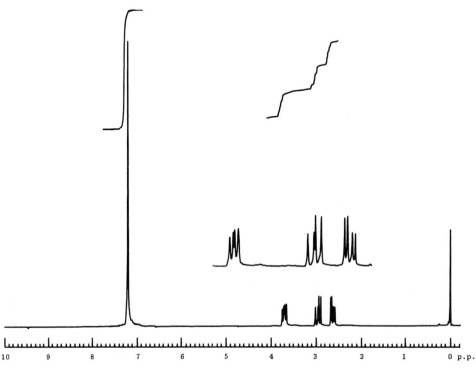

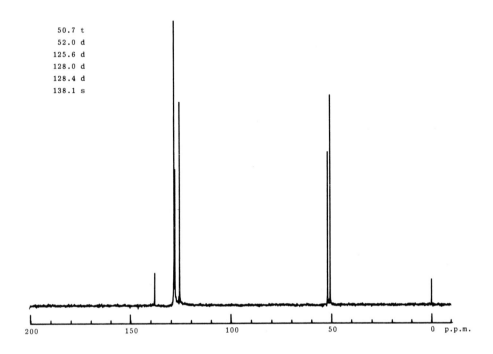

50.7 t
52.0 d
125.6 d
128.0 d
128.4 d
138.1 s

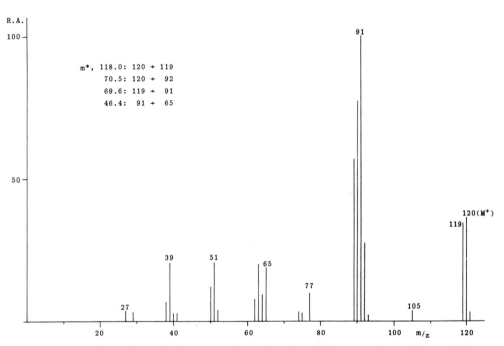

m*, 118.0: 120 → 119
70.5: 120 → 92
69.6: 119 → 91
46.4: 91 → 65

Problem 39   C, 77.4; H, 9.7%

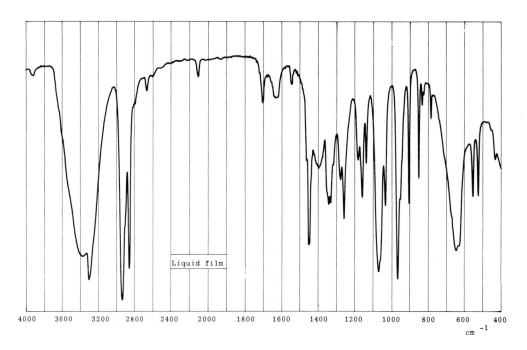

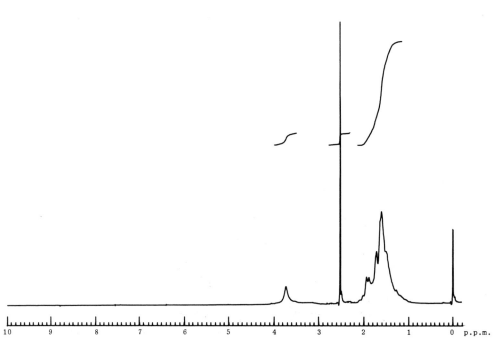

23.5 t
25.6 t
40.1 t
68.7 s
72.8 d
88.4 s

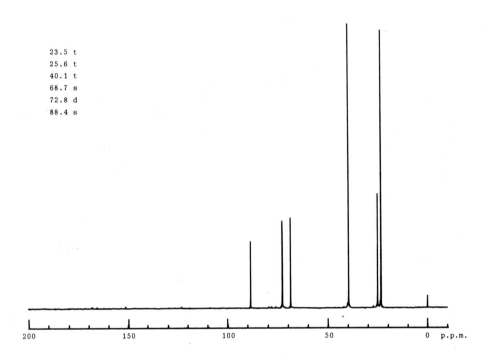

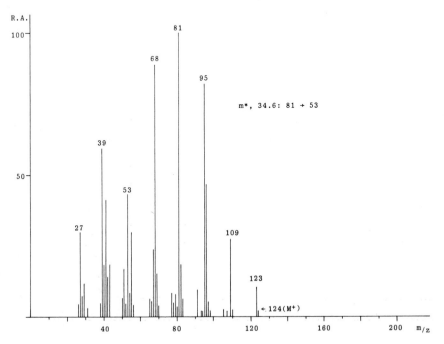

Problem 40   C, 47.4; H, 2.5%

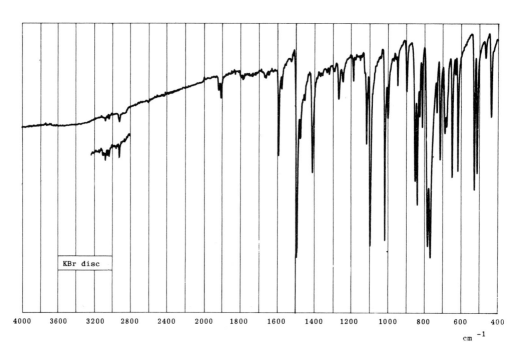

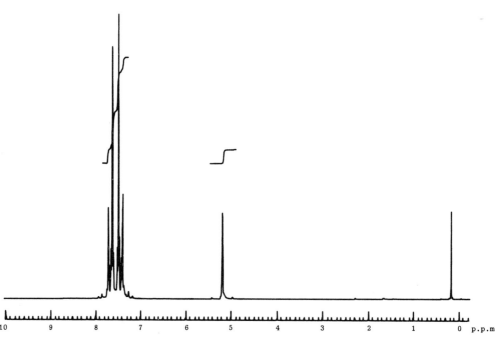

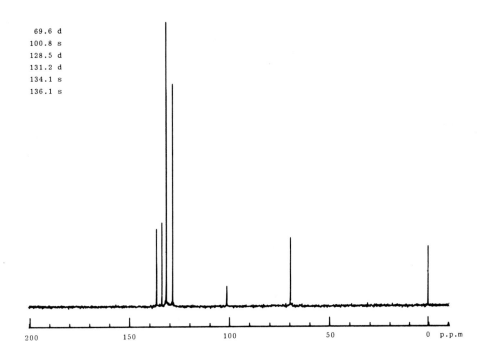

```
69.6 d
100.8 s
128.5 d
131.2 d
134.1 s
136.1 s
```

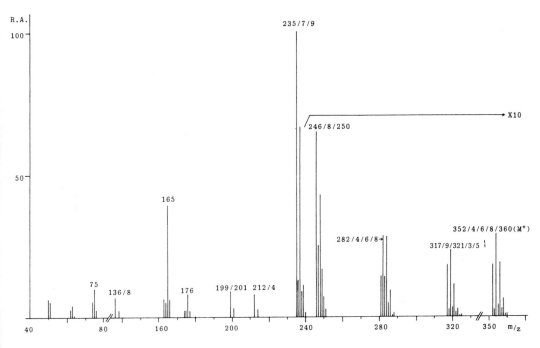

Problem 41 $C_6H_8O_2$

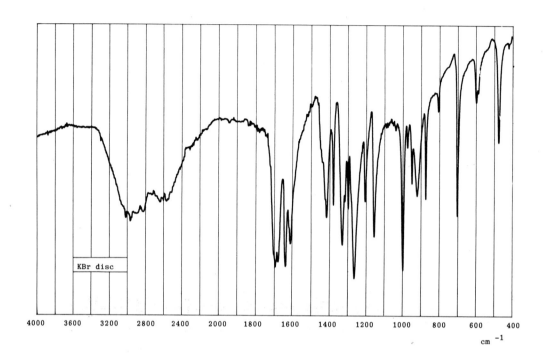

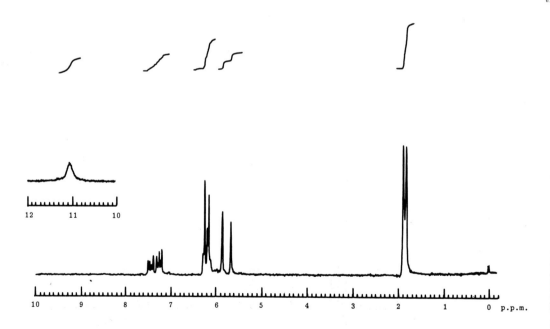

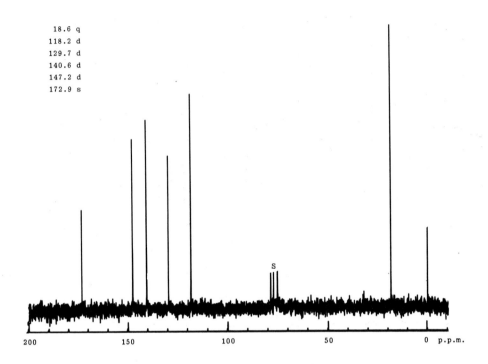

18.6 q
118.2 d
129.7 d
140.6 d
147.2 d
172.9 s

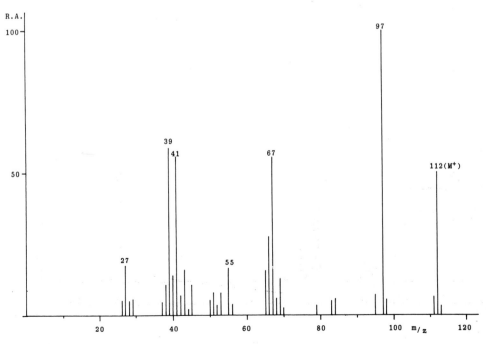

Problem 42 $C_8H_3NO_5$

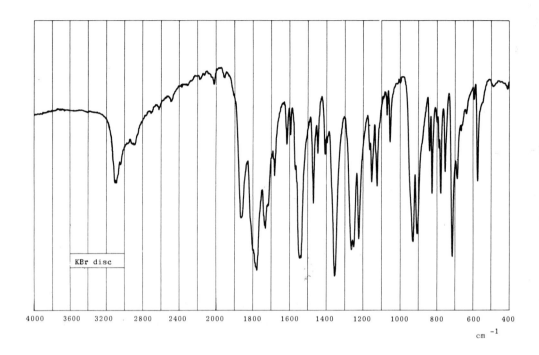

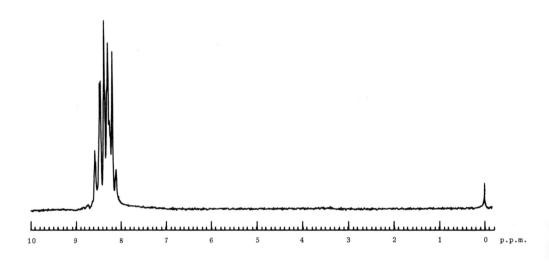

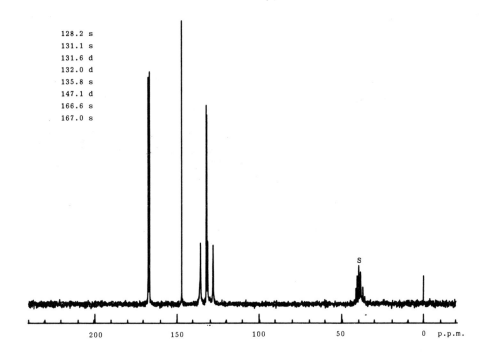

128.2 s
131.1 s
131.6 d
132.0 d
135.8 s
147.1 d
166.6 s
167.0 s

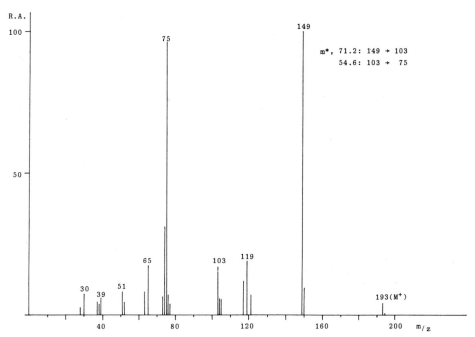

m*, 71.2: 149 → 103
54.6: 103 → 75

Problem 43 $C_6H_6N_2O$

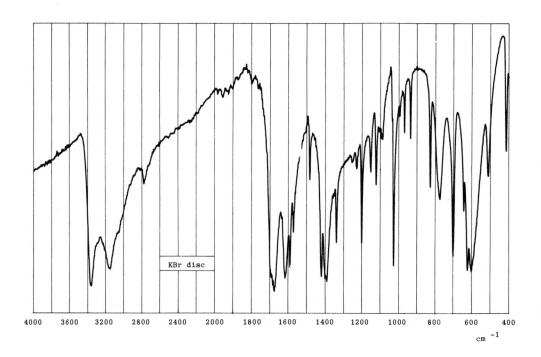

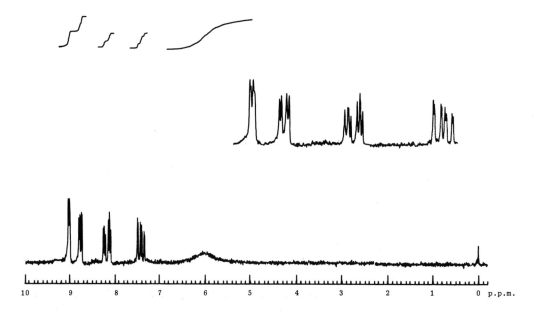

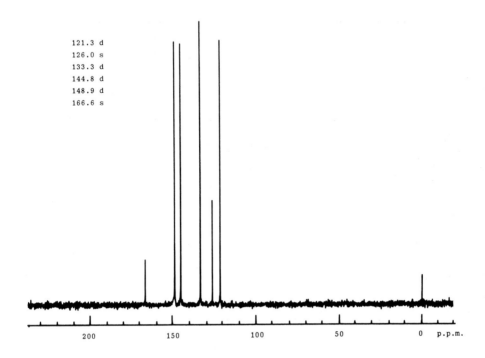

121.3 d
126.0 s
133.3 d
144.8 d
148.9 d
166.6 s

200   150   100   50   0   p.p.m.

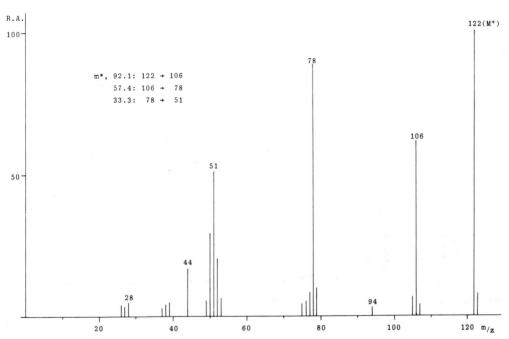

R.A.

100

m*, 92.1: 122 → 106
    57.4: 106 → 78
    33.3: 78 → 51

78

106

122(M⁺)

51

50

44

28

94

20   40   60   80   100   120   m/z

Problem 44 $C_{18}H_{16}$

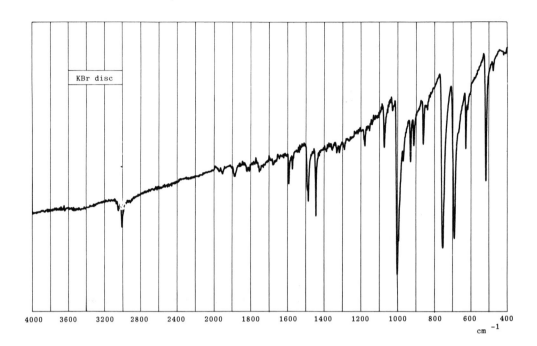

KBr disc

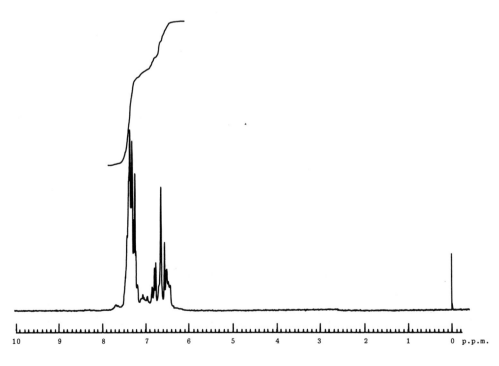

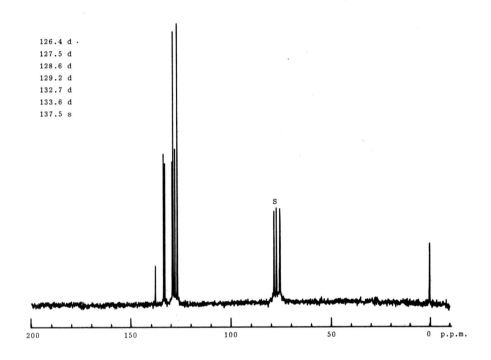

126.4 d ·
127.5 d
128.6 d
129.2 d
132.7 d
133.6 d
137.5 s

S

200   150   100   50   0   p.p.m.

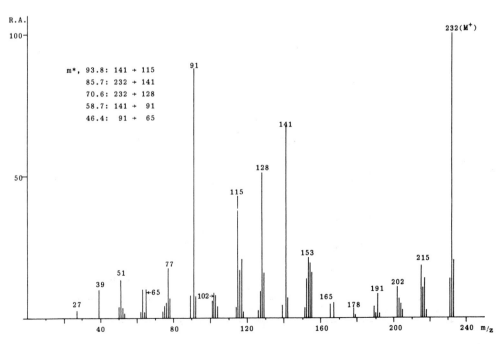

R.A.

100—

m*, 93.8: 141 → 115
    85.7: 232 → 141
    70.6: 232 → 128
    58.7: 141 → 91
    46.4: 91 → 65

91

232(M⁺)

141

128

115

153

50—

77

215

51

39

202

191

|←65

102→|

165

178

27

40   80   120   160   200   240   m/z

Problem 45 $C_8H_{11}NO$

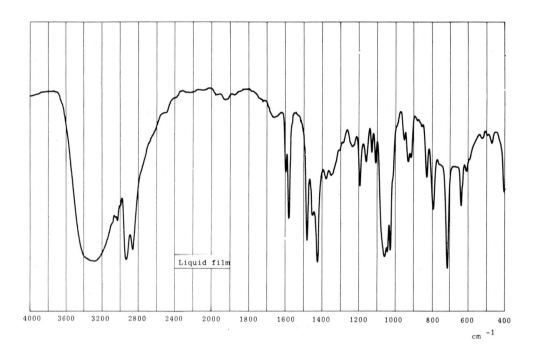

Liquid film

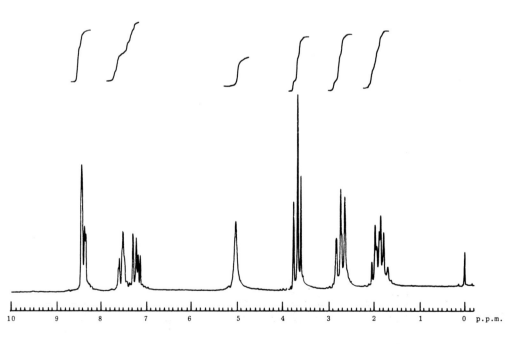

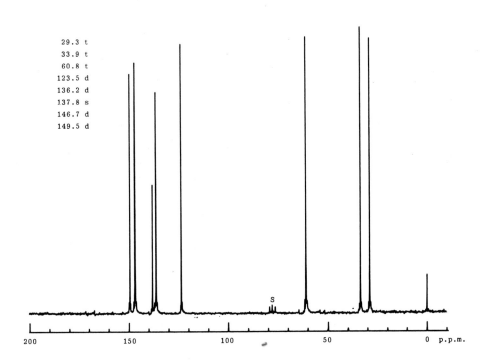

29.3 t
33.9 t
60.8 t
123.5 d
136.2 d
137.8 s
146.7 d
149.5 d

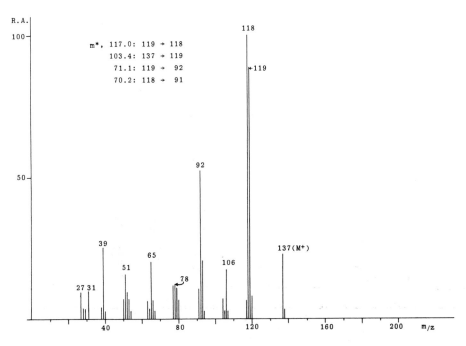

m*, 117.0: 119 → 118
103.4: 137 → 119
71.1: 119 → 92
70.2: 118 → 91

G

Problem 46 C_8H_9N

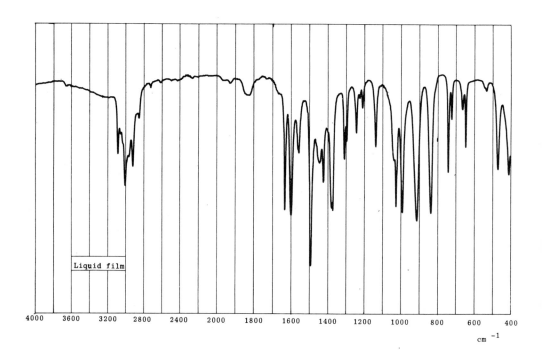

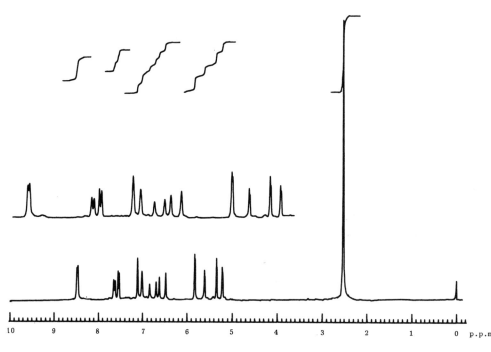

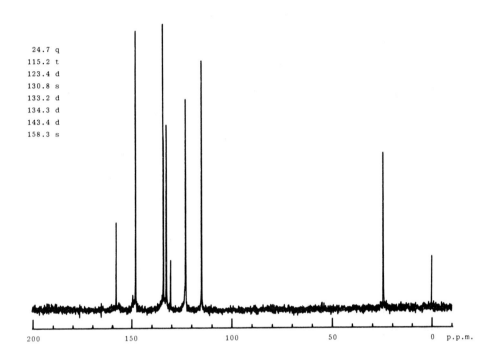

```
24.7 q
115.2 t
123.4 d
130.8 s
133.2 d
134.3 d
143.4 d
158.3 s
```

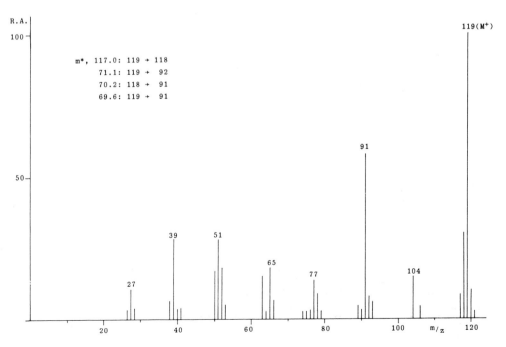

```
m*, 117.0: 119 → 118
     71.1: 119 → 92
     70.2: 118 → 91
     69.6: 119 → 91
```

Problem 47 $C_{10}H_{16}$

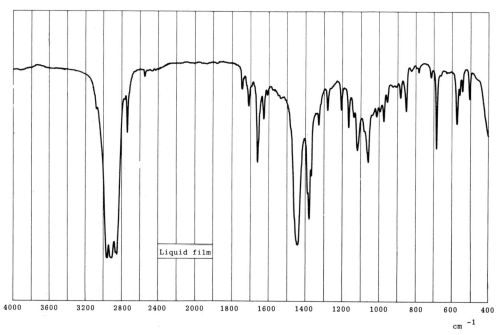

Liquid film

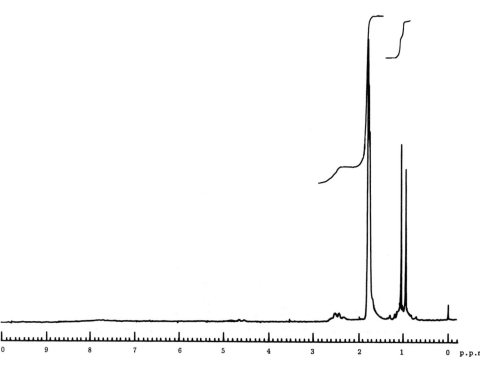

11.0 q
11.5 q
14.1 q
51.7 d
134.2 s
137.7 s

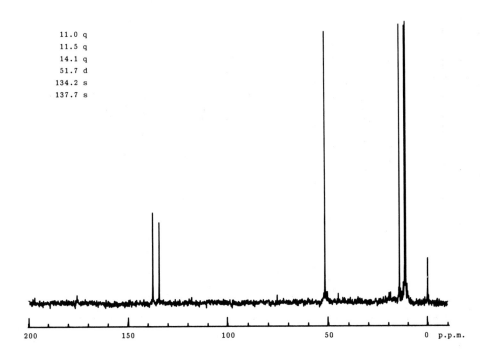

R.A.

100 —

m*, 107.7: 136 → 121
     101.0: 105 → 103
      91.1: 121 → 105
      89.0: 93 → 91
      71.5: 121 → 93
      59.4: 105 → 79

50 —

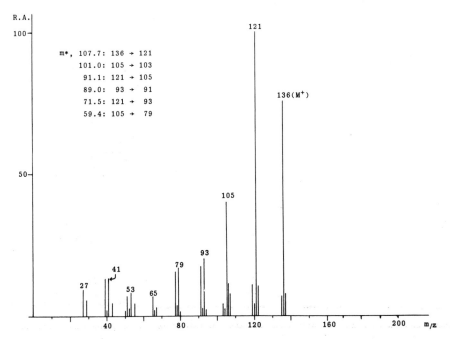

Problem 48 $C_{11}H_{18}O_4$

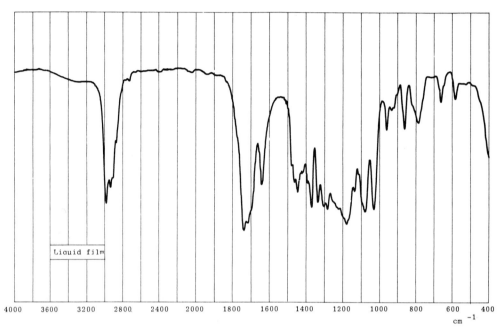

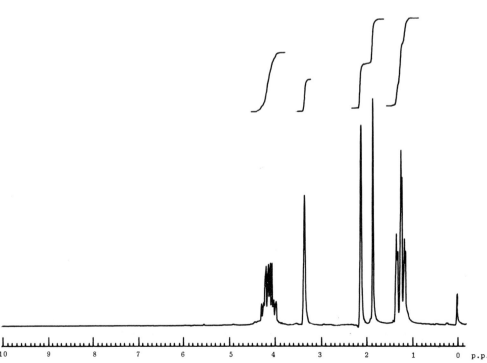

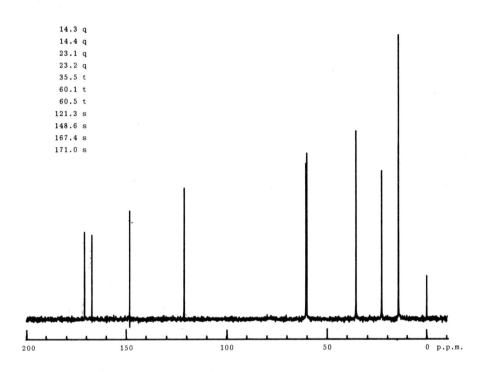

```
14.3 q
14.4 q
23.1 q
23.2 q
35.5 t
60.1 t
60.5 t
121.3 s
148.6 s
167.4 s
171.0 s
```

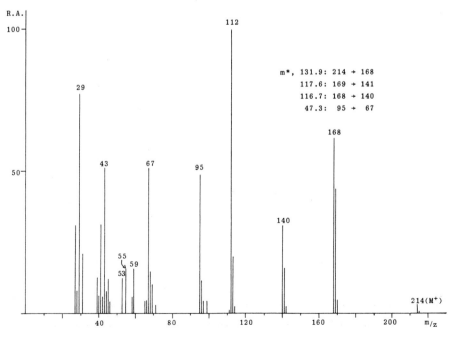

```
m*, 131.9: 214 → 168
    117.6: 169 → 141
    116.7: 168 → 140
     47.3:  95 →  67
```

Problem 49 $C_{14}H_{18}N_2$

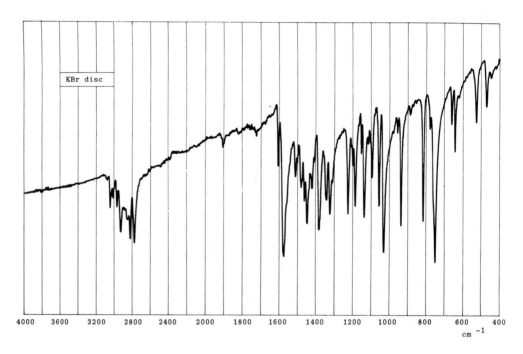

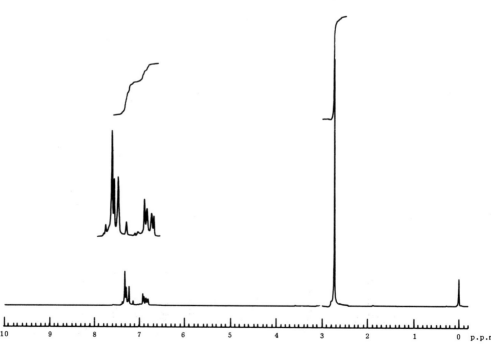

44.3 q
112.7 d
120.7 s
121.8 d
125.3 d
138.8 s
150.6 s

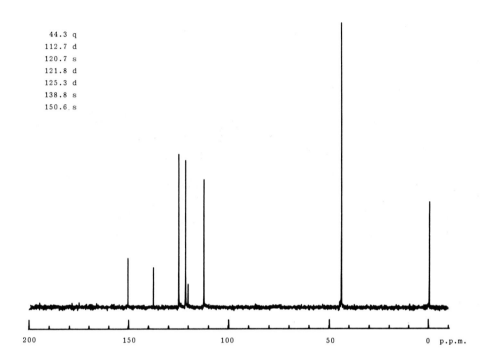

200   150   100   50   0 p.p.m.

R.A.
100 —

168

m*, 145.2: 199 → 170
135.0: 214 → 170

214(M⁺)

183

170

127

154

199

42

44

77

115

107

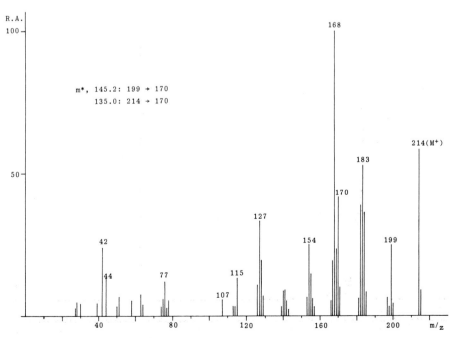

40   80   120   160   200   m/z

Problem 50 $C_{12}H_{14}O_4$

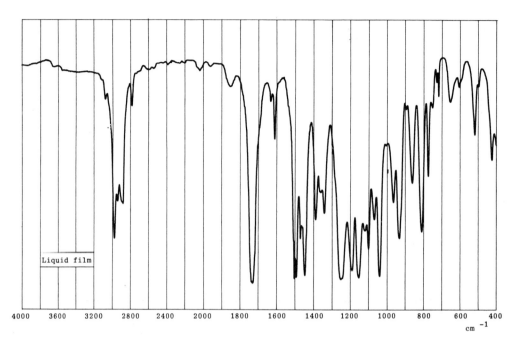

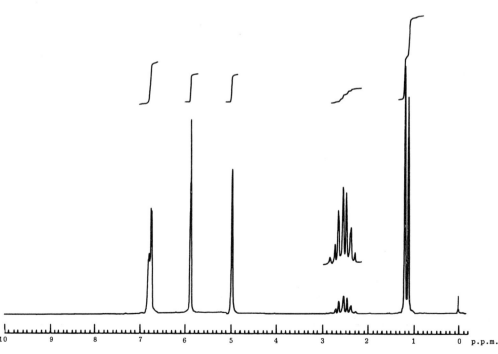

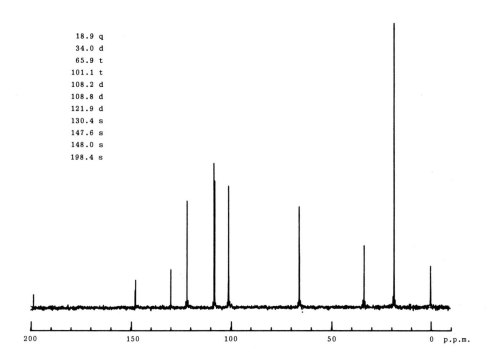

18.9 q
34.0 d
65.9 t
101.1 t
108.2 d
108.8 d
121.9 d
130.4 s
147.6 s
148.0 s
198.4 s

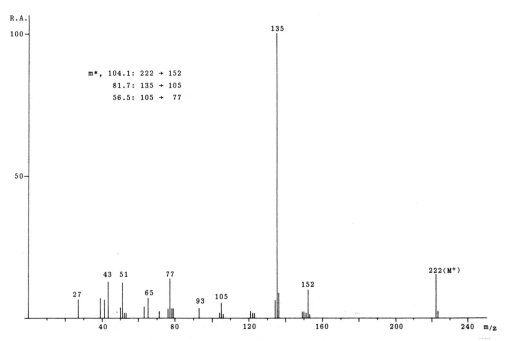

m*, 104.1: 222 → 152
81.7: 135 → 105
56.5: 105 → 77

Problem 51 $C_{10}H_9NO_2$
A substituted benzoic acid.

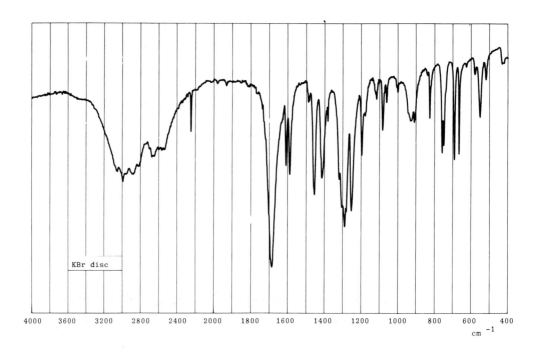

KBr disc

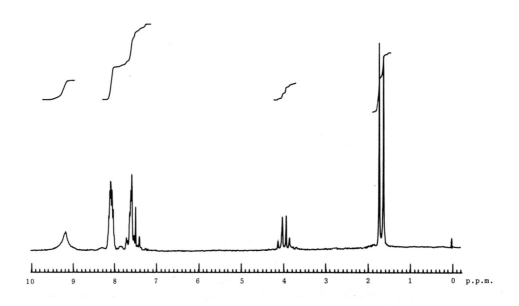

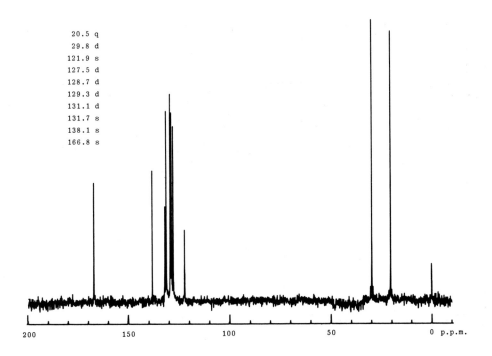

20.5 q
29.8 d
121.9 s
127.5 d
128.7 d
129.3 d
131.1 d
131.7 s
138.1 s
166.8 s

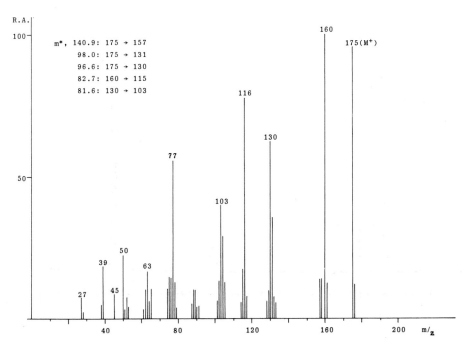

m*, 140.9: 175 → 157
98.0: 175 → 131
96.6: 175 → 130
82.7: 160 → 115
81.6: 130 → 103

Problem 52 $C_6H_{10}O_3$

A mixture of tautomers.

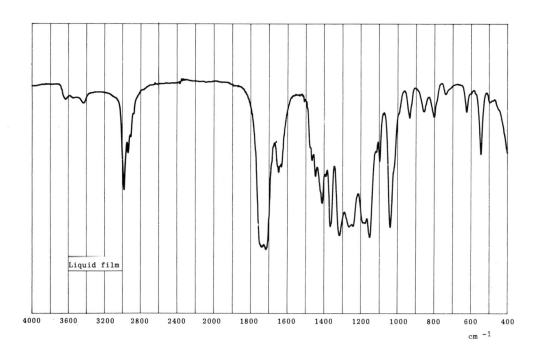

Liquid film

cm^{-1}

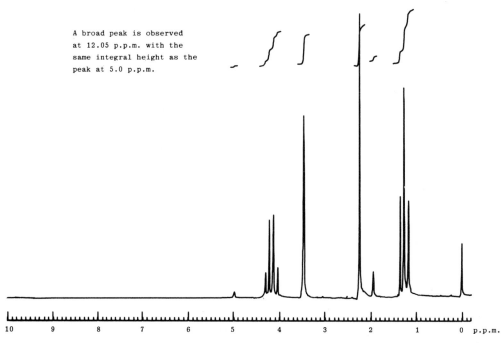

A broad peak is observed
at 12.05 p.p.m. with the
same integral height as the
peak at 5.0 p.p.m.

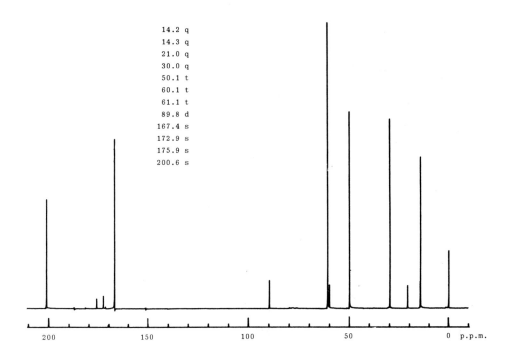

14.2 q
14.3 q
21.0 q
30.0 q
50.1 t
60.1 t
61.1 t
89.8 d
167.4 s
172.9 s
175.9 s
200.6 s

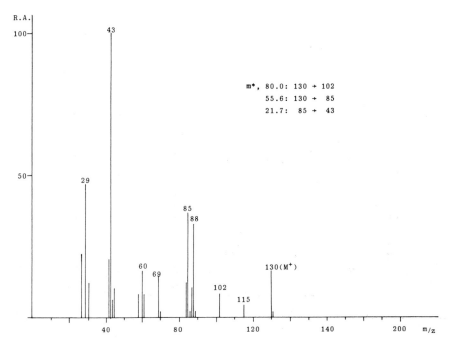

m*, 80.0: 130 → 102
55.6: 130 → 85
21.7: 85 → 43

Problem 53   A derivative of compound 9.

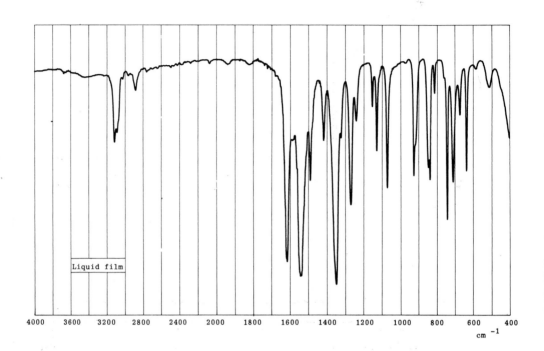

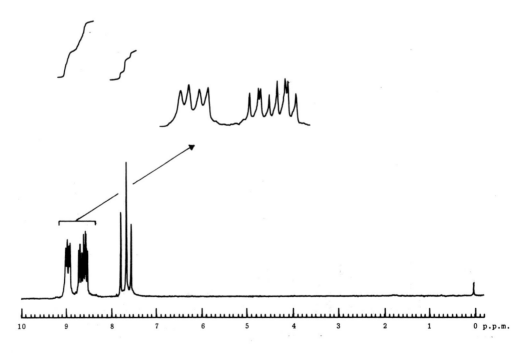

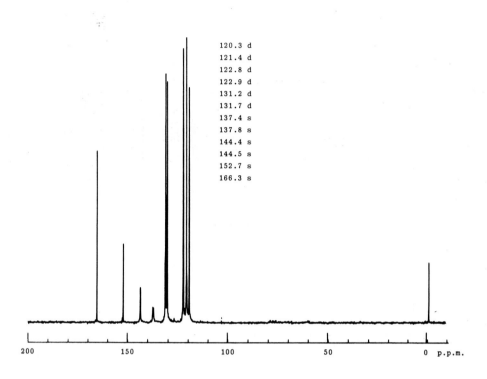

120.3 d
121.4 d
122.8 d
122.9 d
131.2 d
131.7 d
137.4 s
137.8 s
144.4 s
144.5 s
152.7 s
166.3 s

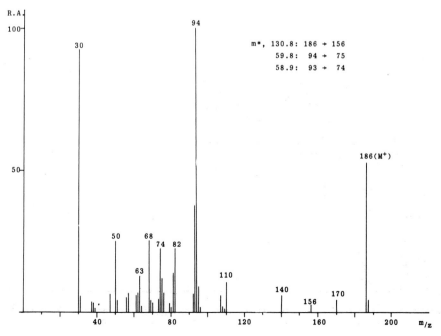

m*, 130.8: 186 → 156
59.8: 94 → 75
58.9: 93 → 74

H

Problem 54 $C_{21}H_{18}O$

A 9-substituted fluorene.

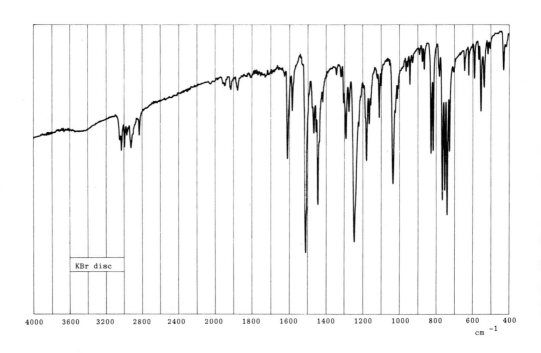

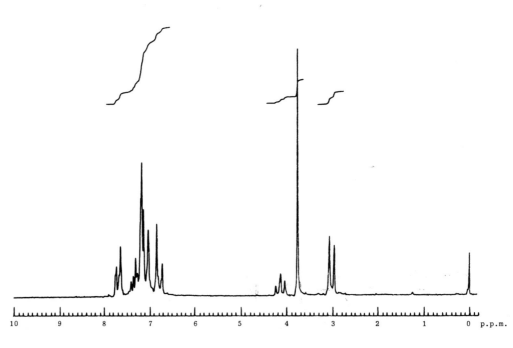

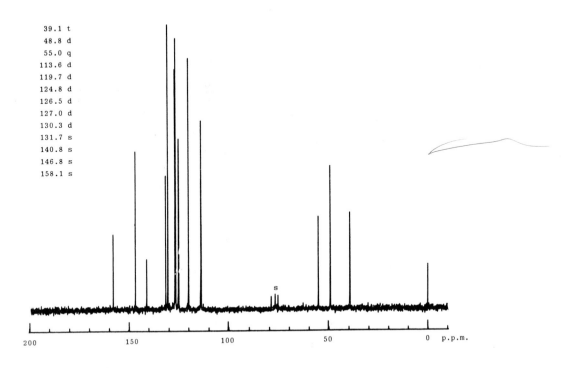

```
39.1 t
48.8 d
55.0 q
113.6 d
119.7 d
124.8 d
126.5 d
127.0 d
130.3 d
131.7 s
140.8 s
146.8 s
158.1 s
```

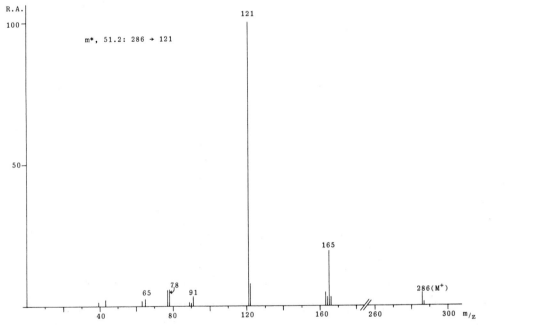

m*, 51.2: 286 → 121

Problem 55 $C_{10}H_{14}N_2$

A naturally occurring 2-substituted-N-alkyl-pyrrolidine.

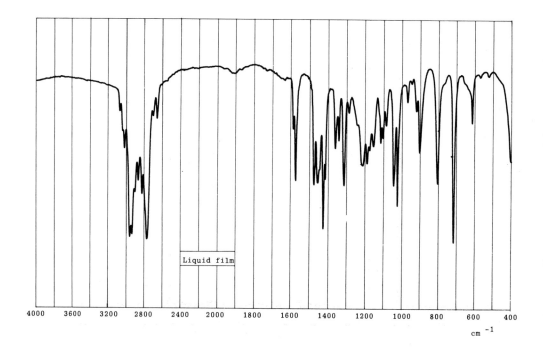

Liquid film

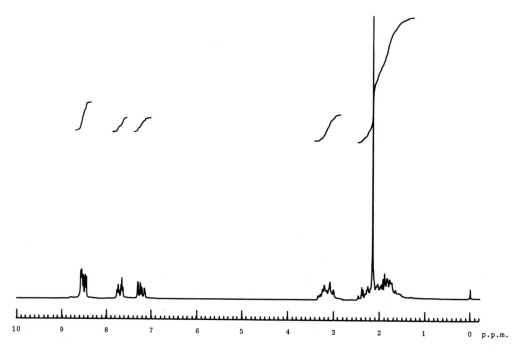

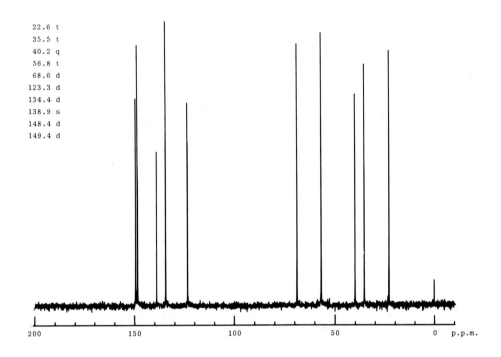

22.6 t
35.5 t
40.2 q
56.8 t
68.6 d
123.3 d
134.4 d
138.9 s
148.4 d
149.4 d

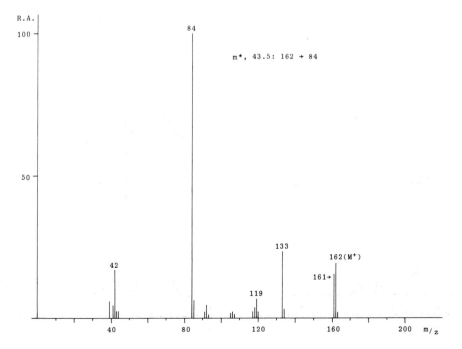

m*, 43.5: 162 → 84

Problem 56 $C_{10}H_{18}O$

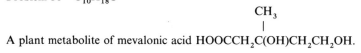

A plant metabolite of mevalonic acid $HOOCCH_2C(OH)CH_2CH_2OH$.

$$CH_3$$

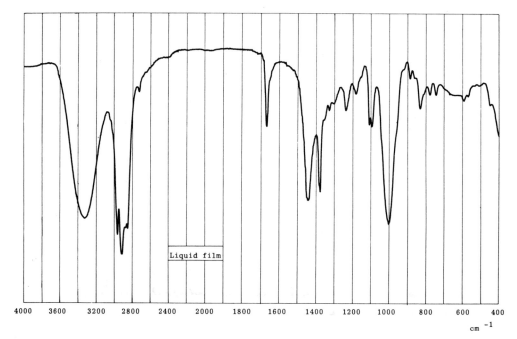

Liquid film

4000   3600   3200   2800   2400   2000   1800   1600   1400   1200   1000   800   600   400

cm^{-1}

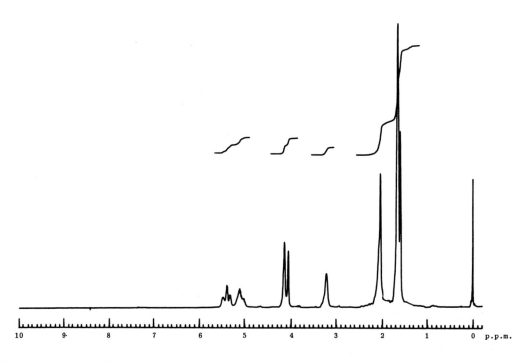

10   9·   8   7   6   5   4   3   2   1   0   p.p.m.

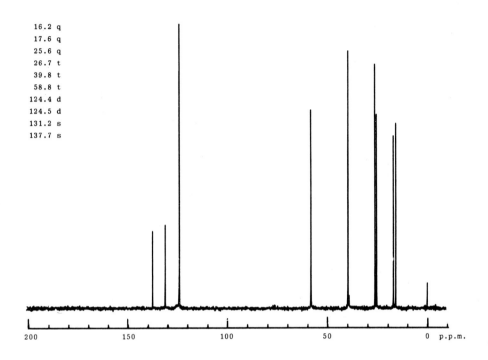

16.2 q
17.6 q
25.6 q
26.7 t
39.8 t
58.8 t
124.4 d
124.5 d
131.2 s
137.7 s

200   150   100   50   0 p.p.m.

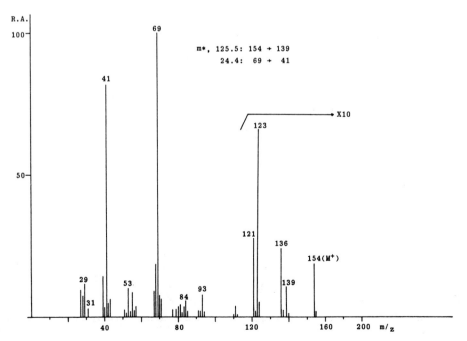

R.A.

100

69

m*, 125.5: 154 → 139
      24.4: 69 → 41

41

X10

123

50

121

136

154(M⁺)

29
31
53
84
93
139

40   80   120   160   200 m/z

Correlation tables

Mass Spectral Losses

Ion	Loss of	Deduction
M-1	H	
M-2	H_2	
M-14	CH_2	Homologue
M-15	CH_3	
M-16	O	$ArNO_2$
	NH_2	$ArCONH_2$
M-17	OH	ArCOOH
M-18	H_2O	Ketone, aldehyde, alcohol, acid
M-26	C_2H_2	Aromatic hydrocarbon
M-27	HCN	N-heteroaromatic, ArCN, $ArNH_2$
M-28	C_2H_4	Et ester, ArOEt, n-Pr ketone
M-29	C_2H_5	Ethyl ketone
	CHO	ArCHO
M-30	CH_2O	$ArOCH_3$
	NO	$ArNO_2$
	C_2H_6	
M-31	CH_3O	Methyl ester
M-32	CH_3OH	Methyl ester
M-41	C_3H_5	Propyl ester
M-42	C_3H_6	Butyl ketone
	CH_2CO	ArNHCOMe, Methyl ketone
M-43	C_3H_7	Propyl ketone
	CH_3CO	Methyl ketone
M-44	C_3H_8	
	CO_2	Ester, anhydride
M-45	CO_2H	Carboxylic acid
	C_2H_5O	Ethyl ester
M-46	C_2H_5OH	Ethyl ester
	NO_2	$ArNO_2$
M-55	C_4H_7	Butyl ester
M-56	C_4H_8	Pentyl ketone, ArOBu
M-57	C_2H_5-CO	Ethyl ketone
	C_4H_9	Butyl ketone
M-58	CH_3COCH_3	Aliphatic methyl ketone
M-60	CH_3COOH	Acetate

Mass Spectral Ions

	Possible formulae	Possible compound type
29	C_2H_5, CHO	Ethyl group, aromatic aldehyde
30	NO	Aromatic and aliphatic nitro compounds
35/37	^{35}Cl, ^{37}Cl	Chloro compounds
36/38	$H^{35}Cl$, $H^{37}Cl$	Chloro compounds
39	C_3H_3	Aromatic and heterocyclic compounds
44	$CH_3CH=NH_2$	Amines
	NH_2CO	Amides
45	$CH_3-CH=OH$	Alcohols
	$CH_3O=CH_2$	Ethers
	COOH	Acids
46	NO_2	Nitro compound
58	$CH_2=C(OH)-CH_3$	Ketones
	$(CH_3)_2N=CH_2$	Amines
59	$C_2H_5-CH=OH$	Alcohols
	$CH_2=C(OH)-NH_2$	Amides
	$C_2H_5O=CH_2$	Ethers
60	$CH_2=C(OH)_2$	Aliphatic acids
65	C_5H_5	Aromatic compounds
71	C_3H_7CO	Propyl ketone
72	$C_4H_{10}N$	Amines
73	$C_3H_7-CH=OH$	Alcohols
	$C_3H_7-O=CH_2$	Ethers
	$CH_2=CH-C(OH)=OH$	Aliphatic acids
74	$CH_2=C(OH)-OCH_3$	Methyl ester of aliphatic acids
76	C_6H_4	Benzene derivatives
77	C_6H_5	Monosubstituted benzene derivative
78	C_6H_6	Aromatic compounds
79/81	C_6H_7, $^{79}Br/^{81}Br$	Aromatic or bromo compound
89	C_7H_5	N and O containing heterocyclics
90	C_7H_6	N and O containing heterocyclics
91	C_7H_7	Aromatic compounds
92	C_6H_6N	Monoalkylpyridine
105	C_6H_5CO	Phenylketone, benzoate ester
121	C_8H_9O	$CH_3OC_6H_4CH_2^-$
122	$C_7H_6O_2$	Alkyl benzoate
123	$C_7H_7O_2$	Alkyl benzoate

Infrared Bands

Bond	Assignment	4000	3000	2000	1600	1200	800	400 cm^{-1}
O-H	Phenols, alcohols							
	Free	m ☐						
	H-bonded	m ☐						
	Carboxylic acids	m ☐						
N-H	Amides, primary	m ☐			☐ m-s			
	and secondary amines							
C-H	Aromatic							
	(stretch)		s ☐					
	(out-of-plane bend)						☐ s	
	Alkanes (stretch)		☐ s					
	CH$_3$- (bend)				m ☐	☐ m		
	-CH$_2$- (bend)				m ☐			
	Alkenes							
	(stretch)		m ☐					
	(out-of-plane bend)						☐ s	
C≡C	Alkyne			m-w ☐				
C≡N	Nitriles			m ☐				
C=O*	Aldehyde				☐ s			
	Ketone				☐ s			
	Acid				☐ s			
	Ester				☐ s			
	Amide				☐ s			
	Anhydride			s ☐ ☐ s				
C=C	Alkene				☐ m-w			
	Aromatic				m-w ☐ ☐ m-w			
C-O	Esters, ethers,					☐ s		
	Anhydrides, alcohols					☐ s		
	Carboxylic acids					☐ s		
N=O	Nitro (R-NO$_2$)				☐ s	☐ s		
C-Hal.	Fluoride					☐ s		
	Chloride						☐ s	
	Bromide, iodide							☐ s

| | | 4000 | 3000 | 2000 | 1600 | 1200 | 800 | 400 cm^{-1} |

* C=O stretching frequencies are typically lowered by about 20-30 cm^{-1} from the values given when the carbonyl group is conjugated with an aromatic ring or an alkene group.

w = weak   m = medium   s = strong

Bands for Out-of-Place Bending Vibrations of Substituted Alkenes

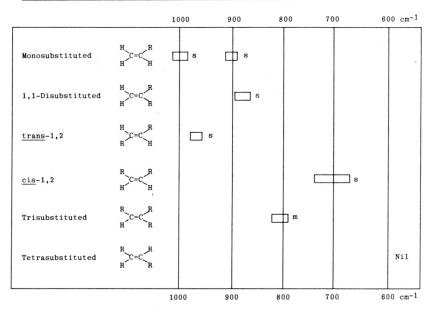

Bands for Out-of-Plane Bending Vibrations of Substituted Benzenoid Compounds

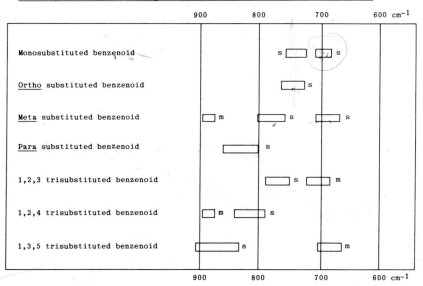

^{13}C Chemical Shifts

Relative to TMS = 0 p.p.m.

	220	200	180	160	140	120	100	80	60	40	20	0 p.p.m.

- $\underline{C}H_3-C\lessgtr$
- $CH_3-N\lessgtr$
- CH_3-O-
- $-\underline{C}H_2-C\lessgtr$
- $-CH_2Br$
- $-CH_2Cl$
- $-CH_2N\lessgtr$
- $-CH_2-O-$
- $\gtrless CH-C\lessgtr$
- $\gtrless CH-Br$
- $\gtrless CH-Cl$
- $\gtrless CH-N\lessgtr$
- $\gtrless CH-O-$
- $\gtrless C-C\lessgtr$
- $\gtrless C-Br$
- $\gtrless C-Cl$
- $\gtrless C-N\lessgtr$
- $C-O-$
- $-C\equiv C-$
- $-C\equiv N$
- $\gtrless C=C\lessgtr$ Benzenes, alkenes
- $\gtrless C=C\lessgtr$ Heteroaromatics
- $\gtrless C=N-$ Heteroaromatics
- $\gtrless C=O$ Anhydrides
- Esters
- Amides
- Acids
- Aldehydes
- Ketones

	220	200	180	160	140	120	100	80	60	40	20	0 p.p.m.

^{1}H Chemical Shifts

Relative to TMS = 0 p.p.m.

	12	11	10	9	8	7	6	5	4	3	2	1	0 p.p.m.
CH$_3$-C$\langle$												▭	
CH$_3$-N$\langle$									▭				
CH$_3$-O-									▭				
CH$_3$-C=C$\langle$										▭			
CH$_3$-C=O										▭			
CH$_3$-Ar										▭			
-CH$_2$-C$\langle$												▭	
-CH$_2$-Halogen								▭▭					
-CH$_2$-O-								▭					
-CH$_2$-N$\langle$								▭▭					
-CH$_2$-C=C$\langle$										▭			
-CH$_2$-C=O										▭			
-CH$_2$-Ar									▭				
$\rangle$CH-C$\langle$												▯	
$\rangle$CH-Halogen							▭▭						
$\rangle$CH-O-								▭▭					
$\rangle$CH-N$\langle$								▭▭					
$\rangle$CH-C=O										▭			
$\rangle$CH-Ar									▭				
H-C≡C-										▭			
H-C=C$\langle$ Alkene, non-conjugated						▭▭							
H-C=C$\langle$ Alkene, conjugated					▭▭								
H-Ar Aromatic				▭▭									
H-Ar Heteroaromatic				▭▭▭									
H-C=O Aldehyde			▭										
ROH Alcohols*						▭▭▭▭							
RCOOH*	▭▭▭												
RNH$_2$*										▭			
R$_2$NH*								▭▭▭▭					

| | 12 | 11 | 10 | 9 | 8 | 7 | 6 | 5 | 4 | 3 | 2 | 1 | 0 p.p.m. |

*Chemical shift dependent upon degree of hydrogen bonding and exchange effects.

Solutions to the problems

The answers are given in the form of references to commonly available listings of organic compounds.

Entries preceded by **A** refer to catalogue numbers in the Numerical Cross Reference List of the 1983–4 edition of the catalogue of the Aldrich Chemical Co. Ltd.

Entries preceded by **B** refer to *Beilstein's Handbuch der Organischen Chemie*. By way of example, **B**, 6(3), 2073 refers to volume 6, 3rd supplement, page 2073.

Entries preceded by **C** refer to the Chemical Abstracts Registry numbers.

Entries preceded by **D** refer to numbers in the *Dictionary of Organic Compounds*, 5th edition (ed. J. Buckingham), Chapman & Hall (New York), 1982.

Entries preceded by **M** refer to numbers in the *Merck Index*, 9th edition.

1.	**A,** C1,962–7.	**B,** 2,194.	**C,** 79–11–8.	**M,** 2083.	
2.	**A,** 18,572–8.	**B,** 9,441.	**C,** 140–29–4.	**D,** in P–00912.	**M,** 1146.
3.	**A,** B2,560–6.	**B,** 12,1019.	**C,** 103–67–3.	**D,** in B–00735.	
4.	**A,** H4,154–4.	**C,** 123–42–2.	**D,** H–02548.	**M,** 2921.	
5.	**A,** D4,740–4.	**B,** 6(3), 2073.	**C,** 128–37–0.	**D,** D–02227.	**M,** 1532.
6.	**A,** 10,847–2.	**B,** 17,33.	**C,** 3141–27–3.	**D,** D–02161.	
7.	**A,** 13,496–1.	**B,** 1,362.	**C,** 108–20–3.	**D,** D–05257.	**M,** 5073.
8.	**A,** 10,128–1.	**B,** 2,256.	**C,** 590–92–1.	**D,** B–03233.	**M,** 1435.
9.	**A,** D19,425–5.	**B,** 5,258.	**C,** 99–65–0.	**D,** D–07341.	**M,** 3269.
10.	**A,** 10,925–8.	**B,** 2,239.	**C,** 554–12–1.	**D,** in P–02371.	**M,** 5982.
11.	**A,** 19713–0.	**B,** 6(3), 695.	**C,** 3307–39–9.		
12.	**A,** M3,265–8.	**B,** 1,392.	**C,** 123–51–3.	**D,** in M–01200.	**M,** 5055.
13.	**A,** 13897–5.	**B,** 1,383.	**C,** 71–41–0.	**D,** P–00492.	**M,** 6916.
14.	**A,** D3,900–2.	**B,** 5,210.	**C,** 583–53–9.	**D,** D–01448.	
15.	**A,** 15,501–2.	**C,** 540–84–1.	**D,** T–04009.	**M,** 5051.	
16.	**A,** 13,170–9.	**B,** 20,159.	**C,** 109–97–7.	**D,** P–02964.	**M,** 7801.
17.	**A,** P2570–1.	**B,** 5(2), 337.	**C,** 1077–16–3.	**D,** P–01225.	
18.	**A,** M785–5.	**B,** 1,736.	**C,** 141–79–7.	**D,** M–02933.	**M,** 5755.
19.	**A,** 23,583–0.	**B,** 13,461.	**C,** 62–44–2.	**D,** in E–00541.	**M,** 59.
20.	**A,** 11,778–1.	**B,** 2(2), 278.	**C,** 108–22–5.	**D,** in P–02391.	**M,** 5064.
21.	**A,** 10,739–5.	**B,** 7,292.	**C,** 122–78–1.	**D,** P–00910.	**M,** 7066.
22.	**A,** V150–3.	**B,** 2(1), 63.	**C,** 108–05–4.	**D,** in E–00514.	**M,** 9644.
23.	**A,** 11,290–9.	**B,** 14,422.	**C,** 94–09–7.	**D,** in A–01024.	**M,** 3691.
24.	**A,** A2,958–5.	**B,** 1,201.	**C,** 106–95–6.	**D,** B–03241.	**M,** 279.
25.	**A,** M6,710–9	**B,** 1,691.	**C,** 108–10–1.	**D,** M–02911.	**M,** 5068.
26.	**A,** 13,376–0.	**B,** 9,573.	**C,** 140–10–3.	**D,** P–01516.	**M,** 2288.

27.	**A,** H1,300–1.	**B,** 1,734.	**C,** 109–49–9.	**D,** H–00882.	
28.	**A,** 11,200–3.	**B,** 9,225.	**C,** 495–69–2.	**D,** H–00979.	**M,** 4591.
29.	**A,** H315–1.	**B,** 1,699.	**C,** 106–35–4.	**D,** H–00322.	
30.	**A,** 18,588–4.	**B,** 4,59.	**C,** 127–19–5.	**D,** D–05432.	**M,** 3214.
31.	**A,** 13,756–1.	**C,** 77–75–8.	**D,** M–02951.	**M,** 5671.	
32.	**A,** 13,827–4.	**C,** 25414–22–6.	**D,** in H–01835.		
33.	**A,** 10,866–9.	**B,** 11,57.	**C,** 98–58–8.	**D,** in B–01931.	**M,** 1406.
34.	**A,** 14,099–6.	**B,** 2,411.	**C,** 623–70–1.	**M,** 2591.	
35.	**A,** D19,850–1.	**B,** 6,251.	**C,** 51–28–5.	**D,** D–07582.	**M,** 3277.
36.	**A,** 535–6.	**B,** 8, 31.	**C,** 90–02–8.	**D,** H–01251.	**M,** 8087.
37.	**A,** C8,573–5.	**B,** 6,373.	**C,** 108–39–4.	**D,** M–02976.	**M,** 2570.
38.	**A,** S500–6.	**B,** 17,49.	**C,** 96–09–3.	**D,** P–01403.	
39.	**A,** E5,140–6.	**B,** 6(2), 100.	**C,** 78–27–3.	**D,** E–01237.	**M,** 3816.
40.	**A,** 10,002–1.	**C,** 50–29–3.	**D,** T–02696.	**M,** 2822.	
41.	**A,** H830–7.	**B,** 2,483.	**C,** 110–44–1.	**M,** 8495.	
42.	**A,** 15,688–4.	**B,** 17,486.	**C,** 641–70–3.		
43.	**A,** N755–9.	**B,** 22,40.	**C,** 98–92–0.	**D,** P–02819.	**M,** 6340.
44.	**A,** D20,800–0.	**B,** 5,691.	**C,** 1720–32–7.		
45.	**A,** P7,120–7.	**C,** 2859–67–8.			
46.	**A,** 12,773–6.	**C,** 140–76–1.			
47.	**A,** 21,402–7.	**C,** 41539–64–4.			
48.	**C,** 42103–98–0.	**D,** in M–01802.			
49.	**A,** 15,849–6.	**C,** 20734–58–1.	**D,** in D–00964.		
50.	**A,** 19,332–1.	**C,** 5461–08–5.			
51.	**A,** 20,980–5.	**C,** 5537–71–3.			
52.	**A,** E964–1.	**B,** 3,632.	**C,** 141–97–9.	**D,** E–00594.	**M,** 3686.
53.	**A,** D19,680–0.	**B,** 5,262.	**C,** 70–34–8.	**D,** F–00383.	**M,** 4057.
54.	**C,** 16305–99–0.				
55.	**A,** 18,637–6.	**C,** 75202–10–7.	**D,** N–00575.	**M,** 6342.	
56.	**A,** 16,333–3.	**B,** 1,457.	**C,** 106–24–1.	**D,** D–06646.	**M,** 4235.